METHUEN'S MONOGRAPHS

ON PHYSICAL SUBJECTS

General Editors:

B. L. WORSNOP, B.SC., PH.D.

G. K. T. CONN, M.A., PH.D.

# SUPERCONDUCTIVITY

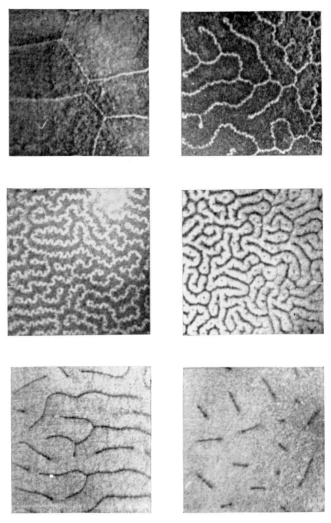

Powder patterns of the intermediate state, showing the shrinking of the superconducting (dark) regions as $h$ takes on the values (left to right, top to bottom) 0, 0·08, 0·27, 0·53, 0·79, and 0·90.

(After Faber, 1958. Reproduced by kind permission of the Royal Society and the author.) *Proc. Roy. Soc. A248, 464, plate 25.*

# Superconductivity

═══

## E. A. Lynton

*Professor of Physics*
*Livingston College, Rutgers University,*
*New Brunswick, New Jersey, U.S.A.*

METHUEN & CO LTD

11 NEW FETTER LANE . LONDON EC4

*First published in* 1962
*Second edition* 1964
*Third edition* 1969
© *E. A. Lynton* 1969
*Printed in Great Britain by*
*Spottiswoode, Ballantyne & Co Ltd*
*London & Colchester*

Distributed in the U.S.A.
by Barnes & Noble, Inc.

QC
612
S8
L9
1969

For Carla

124320

# Acknowledgements

This book has grown, beyond recognition, from a set of lecture notes written and used during my stay at the Institut Fourier of the University of Grenoble in 1959–60. I should like once again to thank my hosts, Professors Néel and Weil and Dr Goodman, for a stimulating and pleasant year. I am very grateful to a large number of people who have helped me with written or oral comments, with news of their unpublished work, with preprints, and with copies of graphs. In particular I thank Drs Coles, Collins, Cooper, Douglass, Faber, Garfunkel, Goodman, Masuda, Olsen, Pippard, Schrieffer, Shapiro, Swihart, Tinkham, Toxen, and Waldram. My colleagues Lindenfeld, McLean, and Weiss provided much helpful discussion. Above all my gratitude is due to Bernard Serin, from whose guidance and friendship I have profited for many years. He found the time to read the entire first draft of the manuscript and suggested many improvements, not all of which I have been wise enough to incorporate.

*September* 1961                                              E. A. LYNTON

# Preface to the Third Edition

This monograph has been brought up to date by changes and additions in all chapters. In addition a complete revision of Chapters V, VI, and VII reflects both the increasingly important role of the Ginzburg-Landau theory and the extensive theoretical and experimental concentration on type II superconductors during recent years. Chapter V has been considerably expanded and now includes – in more logical sequence – the discussion of size effects. Chapter VII is entirely devoted to type II superconductivity, and treats this topic in much greater detail than the previous edition. Further additions to the present edition are the sections on the Josephson effect and on macroscopic coherence in Chapter XI.

In preparing this third edition I have continued to profit from helpful comments of my colleagues at Rutgers University. I am also most grateful to the many preprints sent to me prior to their publication, which made my work much easier.

*September* 1967                                              E. A. LYNTON

# Contents

# Introduction

Although the fascinating phenomenon of superconductivity has been known for fifty years, it is largely through the concentrated experimental and theoretical work of the past decade that a basic (though as yet very incomplete) understanding of the effect has been reached. Far from being an oddity of little physical interest it has been shown to be a co-operative phenomenon of basic importance and with close analogies in a number of fields. At the present time one important period in the development of the subject has been completed, and the next is already well under way, with much effort in theory and experiment to carry our understanding from the general to the particular, from the idealized superconductor to the specific metal. Somewhat coincidentally, there now also is great interest in possible practical applications of superconductivity.

This monograph is a largely descriptive introduction to superconductivity, requiring no more than an undergraduate physics background, and written to serve two functions. It can be a first survey and a stepping stone toward more intensive study for those who intend to become actively engaged in the further development of superconductivity, be it in basic research or in technical applications. Such readers will benefit from the extensive bibliography, listing more than 450 books and articles. At the same time the book is sufficiently complete in its description both of experimental details and of theoretical approaches to be a basic reference for those who wish to be acquainted with the present state of superconductivity. It will enable them to follow further developments as they appear in the scientific and technical literature.

The contents of the book can be grouped into a number of sections which treat the subject of superconductivity in successive layers with increasing resolution of detail. The first three chapters introduce the reader to the principal characteristics of bulk superconductors, and treat these in terms of the basic phenomenological models of London and of Gorter-Casimir. With this section the reader thus acquires a broad outline and a general understanding of the thermodynamic and

1

the static electromagnetic behaviour of idealized, bulk superconductors. The treatment of the subject is then pursued in greater detail along two essentially parallel directions. In the section comprising Chapters IV–VII are discussed those aspects of the behaviour of superconductors which lead to the non-local treatments of Pippard and of Ginzburg and Landau. These more sophisticated phenomenological models account for an interphase surface energy, in terms of which the later chapters of this section describe the intermediate state, phase nucleation, propagation, and supercooling, superconductors of the second kind, and the magnetic behaviour of specimens of small dimensions. Chapters VIII–X can be read without a study of the preceding section (IV–VII) and describe in much detail those characteristics of a superconductor which during the past decade have indicated the microscopic nature of superconductivity, and have led to the theory of Bardeen, Cooper, and Schrieffer. The fundamental aspects of this theory are presented with a minimum of mathematics.

The book closes with a chapter on the behaviour of alloys and compounds, and with one on superconducting devices.

In describing the principal empirical characteristics of superconductors I have tried to include only the key experiments through which the phenomenon in question was established, as well as more recent work which gives the most detailed or the most precise information. It is both unnecessary and impossible in a monograph of this small size to be encyclopaedic either in the enumeration of all pertinent experiments, or in the description of superconducting behaviour in minute detail. My selection of what aspects of the latter to emphasize may appear arbitrary, especially to those whose work has been slighted. The choice was not a judgement of the scientific value of such work, but rather of its didactic usefulness in illuminating the elementary characteristics of superconductors.

# CHAPTER I

# Basic Characteristics

## 1.1. Perfect conductivity and critical magnetic field

The behaviour of electrical resistivity was among the first problems investigated by Kamerlingh Onnes after he had achieved the lique-faction of helium. In 1911, measuring the resistance of a mercury sample as a function of temperature, he found that at about 4°K the resistance falls abruptly to a value which Onnes' best efforts could not distinguish from zero. This extraordinary phenomemon he called *superconductivity*, and the temperature at which it appears the critical temperature, $T_c$ (Kamerlingh Onnes, 1913).

When a metallic ring is exposed to a changing magnetic field, a current will be induced which attempts to maintain the magnetic flux through the ring at a constant value. For a body of resistance $R$ and self-inductance $L$, this induced current will decay as

$$I(t) = I(0)\exp(-Rt/L). \tag{I.1}$$

$I(t)$ can be measured with great precision, for example, by observing the torque exerted by the ring upon another, concentric one which carries a known current. This allows the detection of much smaller resistance than any potentiometric method. A long series of such measurements on superconducting rings and coils by Kamerlingh Onnes and Tuyn (1924), Grassman (1936), and others recently cul-minated in an experiment by Collins (1956), in which a superconduct-ing ring carrying an induced current was kept below $T_c$ for about two and a half years. The absence of any detectable decay of the current during this period allowed Collins to place an upper limit of $10^{-21}$ ohm-cm on the resistivity of the superconductor.† This can be com-pared to the value of $10^{-9}$ ohm-cm for the low temperature resistivity of the purest copper.

There is, therefore, little doubt that a superconductor is indeed a

† Quinn and Ittner (1962) have lowered this upper limit to $10^{-23}$ ohm-cm by looking for the time decay of a current circulating in a thin film tube.

perfect conductor, in the interior of which any slowly varying electric field vanishes. A current induced in a superconducting ring of thickness larger than a few hundred Å will persist an immeasurably long time without dissipation. For wire of smaller diameter Little (1967) has shown that thermodynamic fluctuations will cause a finite lifetime for the decay of the persistent current. The existence of such fluctuations has been demonstrated by Parks and Goff (1967).

Below $T_c$, the superconducting behaviour can be quenched and normal conductivity restored by the application of an external mag-

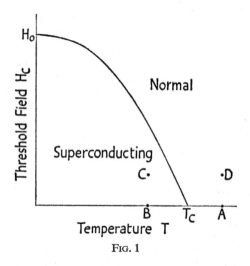

Fig. 1

netic field. This field, $H_c$, is called the *critical* or *threshold magnetic field*, and, as shown in Figure 1, it varies approximately as

$$H_c \approx H_0\left[1 - \left(\frac{T}{T_c}\right)^2\right], \tag{I.2}$$

where $H_0 = H_c$ at $T = 0°K$. It is convenient to introduce reduced coordinates $t \equiv T/T_c$, and $h(t) \equiv H_c(T)/H_0$, in terms of which

$$h \approx 1 - t^2. \tag{I.2a}$$

The actual temperature variation of $h$ is more accurately represented by a polynomial in which the coefficient of the $t^2$ term differs from unity by a few per cent.

The superconductivity of a wire or film carrying a current can be quenched when this reaches a critical value. For specimens sufficiently thick so that surface effects can be ignored, the critical current is that which creates at the surface of the specimen a field equal to $H_c$. Smaller samples remain superconducting with much higher currents than those calculated from this criterion, which is called Silsbee's rule (Silsbee, 1916).

## 1.2. Superconducting elements and compounds

Table I lists all presently known superconducting elements and their

TABLE I

| Element | $T_c$ (°K) | $H_0$ (gauss) |
|---|---|---|
| Aluminium | 1·19 | 99 |
| Cadmium | 0·56 | 30 |
| Gallium | 1·09 | 51 |
| Indium | 3·404 | 293 |
| Iridium | 0·14 | ~20 |
| Lanthanum-$\alpha$ | ~5 | — |
| Lanthanum-$\beta$ | 6·0 | 1600 |
| Lead | 7·18 | 803 |
| Mercury-$\alpha$ | 4·153 | 411 |
| Mercury-$\beta$ | 3·95 | 340 |
| Niobium | 9·46 | 1944 |
| Osmium | 0·66 | 65 |
| Protactinium | 1·4 | — |
| Rhenium | 1·698 | 198 |
| Ruthenium | 0·49 | 66 |
| Tantalum | 4·482 | 830 |
| Technetium | 7·75 | 1410 |
| Thallium | 2·39 | 171 |
| Thorium | 1·37 | 162 |
| Tin | 3·722 | 309 |
| Titanium | 0·39 | 100 |
| Tungsten | 0·012 | 1070 |
| Uranium-$\alpha$ | 0·68 | ~2000 |
| Uranium-$\beta$ | 1·80 | — |
| Vanadium | 5·414 | 1370 |
| Zinc | 0·875 | 53 |
| Zirconium | 0·546 | 47 |

(see Roberts (1966) for most references)

characteristic $H_0$ and $T_c$. In addition there have been found by many investigators, in particular by Matthias and co-workers, by Alekseevskii and co-workers, and by Zhdanov and Zhuravlev (see Matthias, 1957; Roberts, 1966), a very large number of alloys and compounds which also become superconducting. Some of these compounds consist of metals, only one of which by itself becomes superconducting, some have constituents of which neither by itself is superconducting, and some even are semiconductors. The possibility of superconductivity in semiconductors and semimetals has been discussed by M. L. Cohen (1964), and both GeTe (Hein *et al.*, 1964)

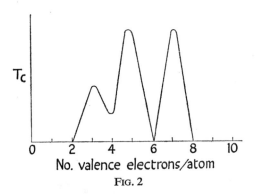

Fig. 2

and $SrTiO_3$ (Schooley *et al.*, 1964, Ambler *et al.*, 1966) have been found to be superconducting at very low temperatures.

The critical temperature of known superconductors range from very low values up to $20 \cdot 05°K$ for a solid solution between $Nb_3Al$ and $Nb_3Ge$ (Matthias *et al.*, 1967). Matthias (1957) has pointed out a number of regularities in the appearance of superconductivity and in the values of $T_c$, the principal of which are the following:

(1) Superconductivity has been observed only for metallic substances for which the number of valence electrons $Z$ lies between about 2 and 8.

(2) In all cases involving transition metals, the variation of $T_c$ with the number of valence electrons shows sharp maxima for $Z = 3, 5.$ and 7, as shown in Figure 2.

(3) For a given value of $Z$, certain crystal structures seem more favourable than others, and in addition $T_c$ increases with a high power of the atomic volume and inversely as the atomic mass.

## 1.3. The Meissner effect, and the reversibility of the S.C. transition

If a perfect conductor were placed in an external magnetic field, no magnetic flux could penetrate the specimen. Induced surface currents would maintain the internal flux, and would persist indefinitely. By

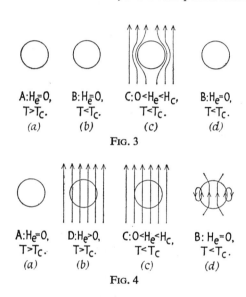

A:$H_{\bar{e}}$0,
$T$>$T_c$.
(a)

B:$H_{\bar{e}}$0,
$T$<$T_c$.
(b)

C:$0$<$H_e$<$H_c$,
$T$<$T_c$.
(c)

B:$H_e$=0,
$T$<$T_c$.
(d)

FIG. 3

A:$H_{\bar{e}}$0,
$T$>$T_c$.
(a)

D:$H_e$>0,
$T$>$T_c$.
(b)

C:$0$<$H_e$<$H_c$,
$T$<$T_c$
(c)

B: $H_e$=0,
$T$<$T_c$.
(d)

FIG. 4

the same token, if a *normal* conductor were in an external field before it became perfectly conducting, the internal flux would be locked in by induced persistent currents even if the external field were removed. Because of this, the transition of a merely perfectly conducting specimen from the normal to the superconducting state would not be reversible, and the final state of the specimen would depend on the path of the transition.

As an example, Figures 3 and 4 show the flux configuration for a perfectly conducting sphere taken from point $A$ in Figure 1 to point $C$

by the different paths *ABC* and *ADC*, respectively. The final field distribution at *C*, as well as that at *B*, depends on whether one proceeded via *B* or via *D*, and the irreversibility of the transition is evident. Careful measurements of the field distribution around a spherical specimen by Meissner and Ochsenfeld (1933), however, indicated that regardless of the path of transition the situation at point *C* is always that shown in Figure 3c: the magnetic flux is expelled from the interior of the superconductor and the magnetic induction *B* vanishes. This is called the *Meissner effect*, and shows that the superconducting transition is reversible.

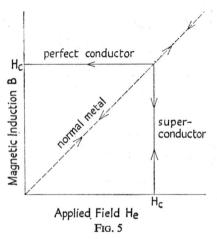

FIG. 5

Figure 5 illustrates this by showing $B$ vs. $H_e$ curves both for a perfect conductor and for a superconductor, taking the case of long cylindrical specimens with axes parallel to the applied field. $H_e$ is a uniform, external field. In increasing field both specimens have $B = 0$ until $H_e = H_c$, when they become normal and $B = H_e$. If the field is now again decreased, the induction inside the perfect conductor is kept at its threshold value $B = H_c$ by surface currents, and in zero field the specimen is left with a net magnetic moment, as is illustrated in Figure 4d. The superconductor, however, expels the flux at the transition and returns reversibly to its initial state with $B = 0$ for $0 < H_e < H_c$.

The vanishing of the magnetic induction, corresponding to the expulsion of the magnetic flux, is the basic characteristic of all ideal superconducting material with dimensions large compared to a basic length which will be mentioned later. It is quite independent of the connectivity of the body, so that if one has a superconductor with a hole, the Meissner effect occurs in the metal and only the hole may be threaded by magnetic flux. The magnetic properties of such a superconducting ring are thus essentially determined by the relative size of the diameter of the ring to the diameter of the hole.

### 1.4. The specific heat

The specific heat of a superconductor consists, like that of a normal metal, of the contribution of the electrons $(C_e)$ and that of the lattice $(C_g)$. For a normal metal at low temperatures

$$C_n = C_{en} + C_{gn} = \gamma T + A(T/\Theta)^3. \tag{I.3}$$

$\gamma$ is the Sommerfeld constant, which is proportional to the density of electronic states at the Fermi surface, $\Theta$ is the Debye temperature, and $A$ a numerical constant for all metals. Experimentally the two contributions to $C_n$ can be separated by plotting $C_n/T$ vs. $T^2$, so that the slope of the resulting curve is $A/\Theta^3$, and the intercept is $\gamma$.

In the superconducting phase

$$C_s = C_{es} + C_{gs}.$$

Figure 6 shows values of both $C_s$ $(H = 0)$ and $C_n$ $(H \geqslant H_c)$ for tin as measured by Corak and Satterthwaite (1954), displaying the characteristic features of a sharp discontinuity in $C_s$ of the order of $2\gamma T_c$ at $T_c$, and a rapid decrease of $C_s$ to values below $C_n$ varying about as $T^3$. It is customary to attribute the difference between $C_s$ and $C_n$ entirely to changes in $C_e$, on the assumption that $C_g$ is the same in both phases. This seems reasonable in view of the electronic nature of the superconducting phenomenon, and is supported by the absence of any observable change in the lattice parameters (Keesom and Kamerlingh Onnes, 1924), and by the detection of only minimal changes in the elastic properties (see, for instance, Alers and Waldorf, 1961). On this assumption

$$C_s - C_n = C_{es} - C_{en}, \tag{I.4}$$

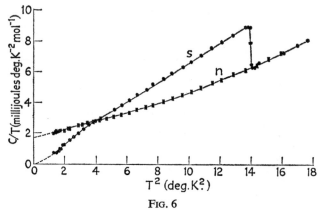

FIG. 6

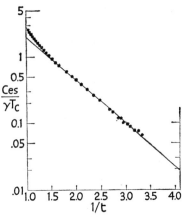

FIG. 7

which allows one to determine $C_{es}$ from measured values of the specific heat difference after $C_{en} = \gamma T$ has been determined separately.

There has recently been some evidence in indium (Bryant and Keesom, 1960; O'Neal *et al.*, 1965) that the part of the specific heat which is proportional to $T^3$ may not be quite equal in the normal and superconducting phases. This could be due to a small inequality of the lattice contributions to the specific heat in the two phases, and indeed Ferrell (1961) has suggested that there may be a shift in the phonon frequency spectrum. However, there is no other evidence to support this. Eliashberg (1962) has proposed alternatively that the difference may be due to a very small $T^3$ contribution to the electron part of the specific heat.

Figure 7 displays $C_{es}$ for tin calculated on the basis of I.4 from the results in Figure 6, plotted logarithmically in units of $1/\gamma T_c$ vs. $1/t$. This shows that for $1/t > 2$, one can represent $C_{es}$ by the equation

$$C_{es}/\gamma T_c = a\exp(-b/t). \tag{I.5}$$

A subsequent chapter will discuss that this is an indication of the existence of a finite gap in the energy spectrum of the electrons separating the ground state from the lowest excited state. The number of electrons thermally excited across this gap varies exponentially with the reciprocal of the temperature. In recent years it has become apparent that such an energy gap determines the thermal properties as well as the high frequency electromagnetic response of all superconductors, and that it must indeed be one of the principal features of a microscopic explanation of superconductivity.

## 1.5. Theoretical treatments

The macroscopic characteristics of a superconductor have been the subject of a number of phenomenological treatments of which the principal ones will be discussed in subsequent chapters. F. and H. London (1935a, b) developed a model for the low frequency electromagnetic behaviour which is based on a point by point relation between the current density and the vector potential associated with a magnetic field. This implies wave functions of the superconducting electrons which even in the presence of such a field extend rigidly to the limits of the superconducting material and then vanish abruptly. A thermodynamic treatment and an associated two-fluid model based on essentially equivalent simplifications were worked out by Gorter and Casimir (1934a, b). These complementary theories provide highly successful and useful tools in the semi-quantitative analysis of many problems involving superconductors. Their limitations become apparent principally in situations in which size and surface effects are important.

Pippard (1950, 1951) has shown that such effects become tractable when one takes into account the finite coherence of the superconducting wave functions which is such as to allow them to vary only slowly over a finite distance. This leads (Pippard, 1953) to a non-local integral relation between the current density at a point and the vector potential in a region surrounding the point. The equation has only been solved for a few special cases. In many instances, however, it reduces to a modified version of the London equation, so that the much simpler London formalism can then be used with the Pippard modifications (Tinkham, 1958).

Ginzburg and Landau (1950) have used a thermodynamic approach to develop an alternate and even more useful method of treating the coherence of the superconducting wave functions. Their treatment explicitly allows for spatial variations in the superconducting order due to size and magnetic field effects, and has come to occupy a very important position in the theory of superconductivity.

A successful microscopic theory of superconductivity has recently been developed by Bardeen, Cooper, and Schrieffer (1957). It is based on the fact, established by Cooper (1956), that in the presence of an attractive interaction the electrons in the neighbourhood of the Fermi surface condense into a state of lower energy in which each electron is paired with one of opposite momentum and spin. Bardeen, Cooper, and Schrieffer (BCS) have been able to show that a finite energy gap separates the state with the largest possible number of Cooper pairs from the state with one pair less. This leads to the correct thermal and electromagnetic properties to display superconductivity.

The attraction between electrons necessary to form Cooper pairs can in principle be due to any suitable kind of interaction. The discovery (Maxwell, 1950; Reynolds *et al.*, 1950) that for many superconducting elements the critical temperature depends on the isotopic mass showed that for these substances the attractive interaction is one between the electrons and the lattice. The BCS theory and its extensions have been worked out on this basis. However, the isotope effect is apparently absent or considerably reduced in some transition metals and their compounds (see Section 8.1). Furthermore the effect of pressure in transition metals does not correlate with the Debye temperatures as it does in non-transition superconductors (Bucher and Olsen, 1964). Kondo (1962) and Garland (1963a, b) have attributed these anomalies to the existence of overlapping bands in the electronic energy spectrum at the Fermi surface. However, there is also a hypothesis that in transition metals the attractive interaction responsible for pairing may be a magnetic one (Matthias, 1960).

# CHAPTER II

# Phenomenological Thermodynamic Treatment

## 2.1. The phase transition

Long before the determination of the reversibility of the superconducting transition by the discovery of the Meissner effect, attempts
had been made to apply thermodynamics to it by Keesom (1924), by
Rutgers (Ehrenfest, 1933), and in particular by Gorter (1933), who
virtually predicted the Meissner effect by pointing out that the success
of these early thermodynamic treatments strongly suggested the
reversibility of the transition.

The discovery of the Meissner effect finally enabled Gorter and
Casimir (1934a) to develop a full treatment of the superconducting
phase transition in a manner analogous to that of other phase transitions. They start with the fact that two phases are in equilibrium with
one another when their Gibbs free energies ($G$) are equal. The free
energy of a superconductor is most easily expressed by a diamagnetic
description developed in Chapter III, which attributes to the superconductor a magnetization $\mathbf{M}$ ($\mathbf{H}_e$) in the presence of an external
field $\mathbf{H}_e$. Then

$$G_s(\mathbf{H}_e) = G_s(0) - \int_0^V dV \int_0^{He} \mathbf{M}(\mathbf{H}_e) d\mathbf{H}_e. \tag{II.1}$$

For an ellipsoid, $\mathbf{M}(\mathbf{H}_e)$ is uniform, and

$$G_s(\mathbf{H}_e) = G_s(0) - V \int_0^{He} \mathbf{M}(\mathbf{H}_e) d\mathbf{H}_e. \tag{II.1'}$$

The last term in this expression gives the work done on the specimen
by the magnetic field. As the magnetization is diamagnetic, that is,
negative, the field raises the energy of the superconducting specimen.

It will be shown in Chapter III that only for a quasi-infinite cylinder
parallel to the external field does the superconducting phase change
into the normal one at a sharply defined value of $H_e$. For all other

shapes, there is an intermediate state consisting of a mixture of normal and superconducting regions. Even under these circumstances, however, any magnetic work is done solely on the superconducting portions, and for any shape of specimen this always equals, per unit volume,

$$\int_0^{H_c} \mathbf{M}(\mathbf{H}_e)\,d\mathbf{H}_e = -H_c^2/8\pi. \tag{II.2}$$

Thus one can write for any specimen:

$$G_s(H_c) = G_s(0) + VH_c^2/8\pi. \tag{II.3}$$

In the normal state the susceptibility is generally vanishingly small, so that

$$G_n(H_c) = G_n(0).$$

Since the condition of equilibrium defining $H_c(T)$ is that

$$G_n(H_c) = G_s(H_c),$$

one has

$$G_n(0) - G_s(0) = VH_c^2/8\pi. \tag{II.4}$$

This is the basic equation of the thermodynamic treatment developed by Gorter and Casimir. As $S = -(\partial G/\partial T)_{p,\,H}$, differentiation of II.4 yields

$$S_n(0) - S_s(0) = -(VH_c/4\pi)(dH_c/dT). \tag{II.5}$$

At $T = T_c$, $H_c = 0$, and $S_n = S_s$. At any lower temperature, $H_c > 0$, and furthermore Figure 1 shows that for $0 < T < T_c$, $dH_c/dT < 0$. Hence the entropies of the two phases are equal at the critical temperature in zero field; at any lower, finite temperature the entropy of the superconducting phase is lower than that of the normal one, indicating that the former is the state of higher order. This ordering will later be shown to follow from a condensation of electrons in momentum space. It follows from Nernst's principle that $S_n = S_s$ at $T = 0$, so that in this limit the slope of the threshold field curve must vanish. As the entropies of the two phases are also equal at $T = T_c$, their difference must pass through a maximum at some intermediate temperature.

Equation II.5 also shows that the latent heat $Q = T(S_n - S_s)$ is zero at the transition in zero field, and is positive when $H_c > 0$. Thus there is an absorption of heat in an isothermal superconducting-to-normal transition, and a corresponding cooling of the specimen when this takes place adiabatically. The resulting possibility of cooling by adiabatic magnetization of a superconductor was suggested by Mendelssohn (Mendelssohn and Moore, 1934) and has been used by Yaqub (1960) for low temperature specific heat measurements of tin.

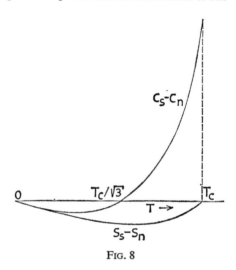

Fig. 8

A further differentiation of II.4 yields, upon multiplication by $T$:

$$C_s - C_n = (VT/4\pi)[H_c(d^2 H_c/dT^2) + (dH_c/dT)^2]. \qquad (\text{II.6})$$

At $T = T_c$, $H_c = 0$, and

$$C_s - C_n = (VT/4\pi)(dH_c/dT)^2_{T = T_c} > 0, \qquad (\text{II.6}')$$

so that the thermodynamic treatment predicts the observed discontinuity in the specific heat. As the entropy difference between the two phases passes through an extremum at some temperature below $T_c$, the specific heats of the two phases at that temperature must be equal, and at even lower temperatures $C_s$ is smaller than $C_n$. Both of course tend toward zero at $T = 0°$. The variation of $C_s - C_n$ as a function of temperature, as well as that of $S_s - S_n$, are shown in Figure 8.

## 2.2. Thermodynamics of mechanical effects

The thermodynamic treatment developed thus far has ignored any changes in the volume at the transition, as well as any dependence of $H_c$ on pressure as well as on temperature. In taking these into account one should begin by considering possible magnetostrictive field effects on the volume in going from II.1 to II.1'. Ignoring this, however, and noting (see Figure 11) that for the special case of a quasi-infinite cylinder parallel to the external field the area under the magnetization curve up to any field value $H_e \leqslant H_c$ is equal to $H_e^2/8\pi$, one can write

$$G_s(H_e) - G_s(0) = (V_s/8\pi) H_e^2. \tag{II.7}$$

Differentiating this with respect to $p$ in order to obtain $V = (\partial G/\partial p)_{T, H}$ yields

$$V_s(H_e) - V_s(0) = (H_e^2/8\pi)(\partial V_s/\partial p)_T. \tag{II.8}$$

Similar differentiation of II.3 and II.4 leads to

$$V_n(H_c) - V_s(0) = \partial/\partial p[V_s H_c^2(T,p)/8\pi],$$

$$V_n(H_c) - V_s(0) = (H_c^2/8\pi)(\partial V_s/\partial p)_T + (V_s H_c/4\pi)(\partial H_c/\partial p)_T. \tag{II.9}$$

Comparing II.9 with II.8 shows that the first term on the right-hand side of the former is just the magnetostriction of the superconductor upon changing the field from zero to the critical value. It is the second term which gives the actual volume change at the transition:

$$V_n(H_c) - V_s(H_c) = (V_s H_c/4\pi)(\partial H_c/\partial p)_T. \tag{II.10}$$

This term exceeds the magnetostrictive one by more than an order of magnitude. The derivatives of II.10 with respect to $T$ and to $p$ yield expressions for the changes at the transition of the coefficient of thermal expansion $\alpha = (1/V)(\partial V/\partial T)$, and of the bulk modulus $\kappa = -V(\partial p/\partial V)$. At $T = T_c$, $H_c = 0$, this yields

$$\alpha_n - \alpha_s = (1/4\pi)(\partial H_c/\partial T)(\partial H_c/\partial p), \tag{II.11}$$

and
$$\kappa_n - \kappa = (\kappa^2/4\pi)(\partial H_c/\partial p)^2. \tag{II.12}$$

There has been extensive experimental work on pressure effects on the critical field. This has been reviewed by Swenson (1960) and summarized most recently by Olsen and Rohrer (1960). These latter authors (1957) and, independently, also Cody (1958), have succeeded in refining earlier work of Lazarev and Sudovstov (1949), and have obtained for different superconducting elements empirical values of the length change of a long rod at the transition. (Andres *et al.*, 1962). Differences in the behaviour of transition and non-transition metals have been pointed out by Bucher and Olsen (1964) (see Section 11.4).

The magnitudes of the several mechanical effects are exceedingly small. Typical values for $\partial H_c / \partial p$ are of the order of $10^{-8}$–$10^{-9}$ gauss/dyne-cm$^{-2}$, and the fractional length change of a long rod is a few parts in $10^{-8}$. Using the above thermodynamic relations this yields a difference in the thermal expansion coefficient of about $10^{-7}$ per degree, and a fractional change in compressibility of one part in $10^5$.

### 2.3. The interrelation between magnetic and thermal properties

One of the most remarkable features of the thermodynamic treatment outlined in the preceding sections is the manner in which it links the magnetic and the thermal properties of a superconductor. Equation II.5, for example, indicates that quite independently of the detailed shape of the magnetic threshold field curve, its negative slope indicates that the superconducting phase has a lower entropy than the normal one. The quantitative verification of an equation such as II.6′, called Rutgers' relation, provides the best available confirmation of the basic reversibility of the superconducting transition. The following table, taken from Mapother (1962), compares for a few particularly favourable elements the specific heat discontinuity measured calorimetrically, with its value calculated with II.6′ from measured threshold field curves. The agreement is seen to be excellent:

| Element | $(C_s - C_n)_{\text{meas}}$ | $(C_s - C_n)_{\text{calc.}}$ |
|---------|------------------------------|-------------------------------|
| | (millijoules/°mole) | |
| Indium | 9·75 | 9·62 |
| Tin | 10·6 | 10·56 |
| Tantalum | 41·5 | 41·6 |

The relations between the thermal properties and the threshold field curve of course also imply that if a specific temperature variation is either assumed or empirically determined for one of the former, this uniquely specifies the temperature variation of the latter. Kok (1934), for example, showed that if one substitutes into equation II.6 a parabolic variation of $H_c$ as given by equation I.2, one obtains a cubic temperature variation of $C_{es}$. It was mentioned in Chapter I that both of these are only fair approximations to the actual temperature dependence of these quantities, and that in fact the threshold field can be represented more accurately by a polynomial which in reduced co-ordinates has the form

$$h(t) = 1 - \sum_{n=2}^{N} a_n t^n. \qquad \text{(II.13)}$$

The first coefficient $a_1$ must vanish, as otherwise $S_s - S_n$ would not vanish at $T = 0$ (see equation II.5), and $\sum_{n} a_n = 1$ to make $h(1) = 0$. If this polynomial is substituted into equation II.6, and one continues to neglect any changes in the lattice specific heat, it follows that

$$\frac{C_{es}}{T} - \frac{C_{en}}{T} = (1/4\pi)(H_0^2/T_c^2)\{(1 - a_2 t^2 - \ldots) \times$$
$$\times (2a_2) + (-2a_2 t - \ldots)\}. \qquad \text{(II.14)}$$

Of the two terms on the left-hand side, the second just equals the Sommerfeld constant $\gamma$. The first is subject to the following general argument: As shown by equation II.5, $S_n > S_s$, and since $S_n$ varies linearly with $T$, $S_s$ must approach zero with some power of $T$ greater than unity. Hence one can write

$$S_{es} \propto T^{1+x}, x > 0,$$

so that

$$C_{es} \propto T^{1+x},$$

and

$$C_{es}/T \propto T^x.$$

It follows, therefore, that no matter what the precise temperature dependence of $C_{es}$ is, $C_{es}/T \to 0$ as $T \to 0°$. Applying this limit to equation II.14 thus yields

$$\gamma = (1/2\pi) a_2 (H_0^2/T_c^2). \qquad \text{(II.15)}$$

An equivalent expression results from applying the above argument directly to equation II.6, and recalling that as $T \to 0$, $dH_c/dT \to 0$. One then obtains

$$\gamma = -(1/4\pi)(H_0^2/T_c^2)(hd^2h/dt^2)_{t=0}. \tag{II.16}$$

Both of these last equations are exact expressions which permit the evaluation of the Sommerfeld constant from a detailed knowledge of the threshold field curve. Mapother (1959, 1962) has carried out a searching analysis of the extent to which magnetic and thermal data can actually be correlated in practice without introducing excessive errors due to extrapolation; Serin (1955) and Swenson (1962) have also discussed the relation between the two types of data.

### 2.4. The Gorter-Casimir two-fluid model
The so-called phenomenological two-fluid models of superconductivity have in common two general assumptions:

(1) The system exhibiting superconductivity possesses an ordered or condensed state, the total energy of which is characterized by an order parameter. This parameter is generally taken to vary from zero at $T = T_c$ to unity at $T = 0°K$, and can thus be taken to indicate that fraction of the total system which finds itself in the superconducting state.

(2) The entire entropy of the system is due to the disorder of non-condensed individual excited particles, the behaviour of which is taken to be similar to that of the equivalent particles in the normal state.

In particular, two-fluid models make the conceptually useful assumption that in the superconducting phase a fraction $\mathscr{W}$ of the conduction electrons are 'superconducting' electrons condensed into an ordered state, while the remaining fraction $1 - \mathscr{W}$ remain 'normal'. The artificiality of this division cannot be overemphasized; its usefulness will presently appear.

The free energy per unit volume of the 'normal' electrons continues to be the same as that of electrons in a normal metal, that is

$$g_n(T) = -\tfrac{1}{2}\gamma T^2 \tag{II.17}$$

where $\gamma$ is the Sommerfeld constant. For the 'superconducting' electrons $g_s(T)$ is taken to be a condensation energy relative to the normal phase, and the considerations of the first section of this chapter show this to be

$$g_s(T) = -H_0^2/8\pi. \tag{II.18}$$

The total free energy per unit volume of the superconducting phase containing a fraction $\mathscr{W}$ of '$s$' electrons and $1 - \mathscr{W}$ of '$n$' electrons is therefore

$$G_s(\mathscr{W}, T) = a(1-\mathscr{W})g_n(T) + b(\mathscr{W})g_s(T). \tag{II.19}$$

The simplest choice of $a(1-\mathscr{W}) = 1 - \mathscr{W}$; $b(\mathscr{W}) = \mathscr{W}$, makes $G_s(\mathscr{W}, T)$ a linear function of $\mathscr{W}$, so that the equilibrium condition $(\partial G/\partial \mathscr{W})_T = 0$ can be satisfied for only one value of $T$ at which $\mathscr{W}$ can assume any value between 0 and 1. This would mean that for any value of $\mathscr{W}$ the normal and superconducting phases can be in equilibrium at only that one temperature, which is not the case. Thus it is necessary to choose $a(1 - \mathscr{W})$ and $b(\mathscr{W})$ with more care. Gorter and Casimir (1934b) chose

$$a(1-\mathscr{W}) = (1-\mathscr{W})^\alpha, \quad b(\mathscr{W}) = \mathscr{W}, \tag{II.20}$$

so that $\quad G_s(\mathscr{W}, T) = -\tfrac{1}{2}(1-\mathscr{W})^\alpha \gamma T^2 - \mathscr{W}H_0^2/8\pi. \tag{II.21}$

Applying the equilibrium condition yields

$$\alpha(1-\mathscr{W})^{\alpha-1} = H_0^2/4\pi\gamma T^2, \tag{II.22}$$

which at $T_c$, with $\mathscr{W} = 0$, reduces to

$$\gamma = (1/4\pi\alpha)(H_0^2/T_c^2). \tag{II.23}$$

Substituting II.23 back into II.22:

$$(1-\mathscr{W})^{\alpha-1} = (T_c/T)^2 = t^{-2}, \tag{II.24}$$

so that $\quad \mathscr{W} = 1 - t^{2/(1-\alpha)}. \tag{II.25}$

Putting this back into II.21 and differentiating to obtain other thermal quantities yields

$$S_s(\mathscr{W}, T) = \gamma T(1-\mathscr{W})^\alpha = \gamma T_c t^{(1+\alpha)/(1-\alpha)}, \tag{II.26}$$

and $\quad C_s(\mathscr{W}, T) = [(1+\alpha)/(1-\alpha)]\gamma T_c t^{(1+\alpha)/(1-\alpha)}. \tag{II.27}$

The value of $\alpha$ must be chosen so as to give a reasonable fit to experimental data. With $\alpha = \tfrac{1}{2}$ one obtains

$$C_s(\mathscr{W}, T) = 3\gamma T_c t^3, \tag{II.27'}$$

$$S_s(T) = \gamma T_c t^3, \tag{II.26'}$$

$$\mathscr{W}(T) = 1 - t^4, \tag{II.25'}$$

and
$$\gamma = (1/2\pi)(H_0^2/T_c^2). \tag{II.23'}$$

Here again is the cubic temperature variation of the specific heat which is only an approximation. Clearly no value of $\alpha$ will change equation II.27 into an exponential expression. A comparison of equation II.23′ with equation II.15 also shows that the choice $\alpha = \frac{1}{2}$ makes $a_2 = 1$, and reduces the polynomial representation of $h(t)$ to a parabolic form. This not only indicates once again the interrelation between the magnetic and thermal properties, but also points up that the Gorter-Casimir model can be used at best only semi-quantitatively. Within this limitation, however, the concept of the two interpenetrating 'fluids' of condensed and uncondensed electrons is very useful in obtaining a semi-quantitative understanding of many superconducting phenomena.

There have been a number of attempts (see [5], p. 280) to improve the quantitative aspects of the Gorter-Casimir model so as to yield more nearly the correct exponential variation of $C_{es}$ and the corresponding non-parabolic dependence of $h(t)$. These modifications have either tried different functional forms for $a(1 - \mathscr{W})$ and $b(\mathscr{W})$ in equation II.19, or have introduced additional adjustable parameters. Some of these variations do yield considerably better equations for the thermal and magnetic superconducting properties. However, the principal virtue of a two-fluid model is to provide a conceptual tool of primarily qualitative nature, and the various suggested improvements rarely add much to the basic physical picture of the two groups of electrons.

The two-fluid model has in recent years been widely used to describe the motion of vortices in type II superconductors (see Section 7.3) and to develop appropriate phenomenological theories. Rickayzen (1966) has attempted to justify this, although serious questions about the applicability of a two-fluid model to vortex motion has since been raised by Caroli and Maki (1967).

# CHAPTER III

# Static Field Description

## 3.1. Perfect diamagnetism

Even in the absence of a microscopic explanation of the phenomenon of superconductivity, it is reasonable to assume that the vanishing of the magnetic induction at the interior of a superconductor is due to induced surface currents.† In the presence of an external magnetic field, the magnitude and distribution of this current is just such as to create an opposing interior field cancelling out the applied one. A formal description of a macroscopic superconductor in the presence of an external field $\mathbf{H}_e$ is, therefore, the following:

*in the interior:* $\mathbf{B}_i = \mathbf{H}_i = \mathbf{M}_i = 0$, where $\mathbf{M}_i$ is the magnetization per unit volume;

*at the surface:* $\mathbf{J}_s \neq 0$, where $\mathbf{J}_s$ is the surface current density; and

*outside:* $\quad\mathbf{B}_e = \mathbf{H}_e + \mathbf{H}_s$, where $\mathbf{H}_s$ is the field due to the surface currents.

It is this field which causes the distorted field distribution near a superconductor as shown in Figure 3c.

Although this description is formally correct, it is much more convenient to replace it by an equivalent one which treats the superconductor in the presence of an external field as a magnetic body with an interior field and magnetization such that

*in the interior:* $\mathbf{B}_i = 0, \mathbf{H}_i \neq 0, \mathbf{M}_i \neq 0$;

*at the surface:* $\mathbf{J}_s = 0$; and

*outside:* $\quad\mathbf{B}_e = \mathbf{H}_e + \mathbf{H}_s$, where now $\mathbf{H}_s$ is the field due to the magnetization of the sample.

† That electron currents and not, for example, spins are responsible for the diamagnetism of a superconductor is demonstrated by its gyromagnetic ratio which is found to have the value of $-e/2mc$ (Kikoin and Goobar, 1940; cf. [1], p. 50 and p. 193; [2], p. 83).

As
$$\mathbf{B} = \mathbf{H} + 4\pi\mathbf{M},$$

this description is equivalent to attributing to the superconductor a magnetization per unit volume

$$\mathbf{M}_i = -(1/4\pi)\mathbf{H}_i \tag{III.1}$$

which means that the superconductor has the ideal diamagnetic susceptibility of $-1/4\pi$.

### 3.2. The influence of geometry and the intermediate state

The great convenience of the diamagnetic mode of description is illustrated by considering an ellipsoidal superconducting specimen in an external field $\mathbf{H}_e$ which is parallel to the major axis. The conventional proof, that inside a uniform ellipsoid $\mathbf{B}$, $\mathbf{H}$, and $\mathbf{M}$ are all constant and parallel to $\mathbf{H}_e$, is independent of susceptibility and therefore applies to the superconductor. Further standard treatments show that (with vector notation now unnecessary):

$$H_i = H_e - 4\pi D M_i, \tag{III.2}$$

where $D$ is the demagnetization coefficient of the specimen. For an ellipsoid of revolution this is given by

$$D = \left(\frac{1}{e^2} - 1\right)\left(\frac{1}{2e}\log\frac{1+e}{1-e} - 1\right).$$

$a$ and $b$ are, respectively, the semi-major and semi-minor axes, and $e \equiv (1 - b^2/a^2)^{1/2}$. For an infinite cylinder with its axis parallel to $H_e$, $D = 0$; for an infinite cylinder transverse to the field, $D = \frac{1}{2}$, and for a sphere, $D = 1/3$.

Combining III.1 and III.2 yields:

$$M_i = -H_e/4\pi(1 - D) \tag{III.3}$$

and

$$H_i = H_e/(1 - D). \tag{III.4}$$

In the neighbourhood of the superconductor, the external field is distorted by the magnetization of the specimen. It follows from the continuity of the normal component of $B$ and of the tangential component of $H$ that for an ellipsoidal specimen the exterior field distribution is as shown in Figure 9. At the equator of the specimen

$$H_{eq} = H_i = H_e/(1-D), \tag{III.5}$$

and at the pole

$$H_p = B_i = 0. \tag{III.6}$$

For the longitudinal infinite cylinder with axis parallel to $H_e$, $D = 0$ and $H_{eq} = H_e$. The exterior field at the surface of the specimen is, therefore, everywhere the same, and the cylinder remains entirely superconducting until the applied field becomes equal to the critical

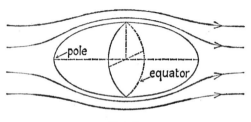

FIG. 9

value $H_c$. The entire body then becomes normal. The magnetization curve for such a specimen is shown in Figure 10, in which for convenience $-4\pi M$ is plotted against $H_e$.

For all other ellipsoidal shapes, $D \neq 0$, and the non-uniformity of the field distribution around the superconductor raises the question of what happens when $H_{eq} = H_c > H_e$. To assume that a portion of the specimen near the equator then becomes normal, as shown in Figure 11, would lead to a contradiction: the boundary between the superconducting and normal regions occurs where $H = H_c$, but in the now normal region the field would equal $H_e < H_c$! There is, in fact, no simple, large-scale division of such a specimen into normal and superconducting regions, which allows a field distribution such that $H \geq H_c$ in the former, $H < H_c$ in the latter, and $H = H_c$ at the boundaries.

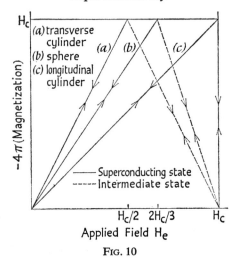

FIG. 10

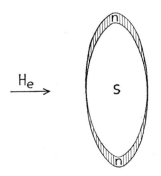

FIG. 11

Instead one must postulate, as was first done by Peierls (1936) and by F. London (1936), that once $H_e \geqslant (1 - D)H_c$, the entire specimen is subdivided into a small-scale arrangement of alternating normal and superconducting regions, with $B = H_c$ in the normal regions, and $B = 0$ in the others. The distribution of these regions varies in such a way that the total magnetization per unit volume changes linearly from

$$M_i = -H_e/4\pi(1-D) = -H_c/4\pi \quad \text{at } H_e = H_c(1-D),$$

to $\quad M_i = 0 \quad\quad\quad\quad\quad\quad\quad\quad \text{at } H_e = H_c.$

Hence, for $(1-D)H_c \leqslant H_e \leqslant H_c$,

$$M_i = -(1/4\pi D)(H_c - H_e), \tag{III.7}$$

$$H_i = H_e - 4\pi D M_i = H_c, \tag{III.8}$$

$$B_i = H_c - (1/D)(H_c - H_e). \tag{III.9}$$

Magnetization curves for a transverse cylinder $(D = \frac{1}{2})$ and for a sphere $(D = 1/3)$ are also shown in Figure 10. Note that the area under each of the curves is given by

$$\int_0^{H_c} M_i dH_e = -H_c^2/8\pi. \tag{III.10}$$

This is just the magnetic work done on the specimen in raising the field from zero to $H_c$, as cited in equation II.2. In the region $(1-D)H_c \leqslant H_e \leqslant H_c$, in which the specimen is neither entirely normal nor entirely superconducting, it is said to be in the *intermediate state*. The detailed structure of this state will be discussed in Chapter VI; at this time it is only necessary to emphasize that this intermediate state exists, in some field interval, for any geometry other than that of a quasi-infinite cylindrical sample parallel to the external field.

### 3.3. Trapped flux
It is important to distinguish the reasons and conditions for the intermediate state from those giving rise to the phenomenon of *trapped flux* or the incomplete Meissner effect. As mentioned earlier, the magnetic flux threading a multiply connected superconductor is trapped by an indefinitely persisting current, and cannot change unless the superconductivity of the specimen is quenched. A similar situation can arise in a simply-connected but non-homogeneous superconductor. Strains, concentration gradients, and other imperfections can create inside a superconductor regions with anomalously

high critical fields. Thus if such a non-ideal specimen is placed in a magnetic field sufficiently high to make it entirely normal, and the field is then reduced, the anomalous regions will become superconducting before the bulk of the specimen. Should some of these regions be multiply-connected, then the flux threading them at the moment of their transition into superconductivity can no longer escape, and is trapped even when the external field is reduced to zero, for as long as the specimen remains superconducting.

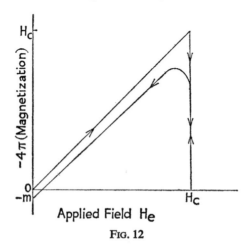

FIG. 12

As a result, after it has once been normal in an external field, such an imperfect specimen is less than perfectly diamagnetic in an external field $H < H_c$, and retains a paramagnetic moment in zero field. This is shown in Figure 12 for a long cylinder parallel to the field, using the same units as in Figure 10. The ratio of $m$ to $-H_c$ is called the fraction of trapped flux.

### 3.4. The perfect conductor

To emphasize once again the difference between a perfect conductor and a superconductor, it is useful to outline an electromagnetic treatment of the former, as developed by Becker *et al.* (1933) just before the discovery of the Meissner effect.

In a perfect conductor, the equation of motion for an electron of mass $m$ and charge $e$ in the presence of an electric field $\mathbf{E}$ does not contain a retarding term and would simply be

$$m\dot{\mathbf{v}} = e\mathbf{E}. \qquad (\text{III}.11)$$

In terms of the current density $\mathbf{J} = ne\mathbf{v}$, where $n$ is the number density of the electrons, one can write III.11 in the form

$$\mathbf{E} = (4\pi\lambda_L^2/c^2)\,\dot{\mathbf{J}}, \qquad (\text{III}.12)$$

where

$$\lambda_L^2 \equiv mc^2/4\pi ne^2. \qquad (\text{III}.13)$$

The parameter $\lambda_L$ has the dimensions of length, and for a density of electrons corresponding to one electron per atom it has a value of the order of $10^{-6}$ cm.

Using Maxwell's equation $\text{curl}\,\mathbf{E} = -\dot{\mathbf{H}}/c$, one finds that

$$(4\pi\lambda_L^2/c)\,\text{curl}\,\dot{\mathbf{J}} + \dot{\mathbf{H}} = 0, \qquad (\text{III}.14)$$

and applying another Maxwell equation $\text{curl}\,\mathbf{H} = 4\pi\mathbf{J}/c$ yields for the perfect conductor the equation

$$\nabla^2\dot{\mathbf{H}} = \dot{\mathbf{H}}/\lambda_L^2. \qquad (\text{III}.15)$$

Von Laue (1949) showed that the solution of III.15 for any specimen geometry yields a value of $\dot{\mathbf{H}}$ which decreases exponentially as one enters the specimen. For a semi-infinite slab extending in the $x$-direction from the plane $x = 0$, the appropriate solution is

$$\dot{\mathbf{H}}(x) = \dot{\mathbf{H}}(0)\exp(-x/\lambda_L). \qquad (\text{III}.16)$$

Clearly, for $x \gg \lambda_L$, $\dot{\mathbf{H}}(x) \approx 0$. Thus equation III.16 confirms that in the interior of a perfect conductor the magnetic field cannot change in time from the value it had when the specimen became perfectly conducting.

### 3.5. The London equations for a superconductor
The incorrectness of III.16 was demonstrated by the discovery of Meissner and Ochsenfeld (1933) that regardless of the magnetic history of the specimen, the field inside a superconductor always vanishes. F. and H. London (1935a, b; see [2]) therefore proposed to

add to Maxwell's equations the following two relations in order to treat the electromagnetic properties of a superconductor:

$$\mathbf{E} = (4\pi\lambda_L^2/c^2\dot{\mathbf{J}}), \tag{A}$$

and

$$(4\pi\lambda_L^2/c)\,\mathrm{curl}\,\mathbf{J} + \mathbf{H} = 0. \tag{B}$$

Replacing the field by a vector potential $\mathrm{curl}\,\mathbf{A} = \mathbf{H}$ and choosing a gauge such that $\mathrm{div}\,\mathbf{A} = 0$, (B) reduces to

$$\frac{4\pi\lambda_L^2}{c}\mathbf{J} + \mathbf{A} = 0. \tag{B'}$$

Note that (A) is identical to III.2, and thus describes the property of perfect conductivity, but that the difference between (B) and III.4 is the important one that application of Maxwell's equations now leads to

$$\nabla^2\mathbf{H} = \mathbf{H}/\lambda_L^2. \tag{III.17}$$

Solution of this for any geometry now shows that $\mathbf{H}$, and not only $\dot{\mathbf{H}}$, decays exponentially upon penetrating into a superconducting specimen. For the semi-infinite slab described above, the solution of III.17 is

$$\mathbf{H}(x) = \mathbf{H}(0)\exp(-x/\lambda_L), \tag{III.18}$$

which shows that for $x \gg \lambda_L$, $\mathbf{H}(x) \approx 0$, in accordance with the Meissner effect.

Clearly the London equations (A) and (B) do not, in fact, yield the complete exclusion of a magnetic field from the interior of a superconductor. Instead, they predict the penetration of a field such that it decays to $1/e$ of its surface value in a distance $\lambda_L$. This is called the London penetration length. Its existence has been fully confirmed experimentally, although empirical values are consistently higher than those predicted by the defining equation III.13, as will be discussed in a later chapter. The existence of this slight penetration of an exterior field must be taken into consideration in the discussion of superconducting thin films, wires, or colloidal particles, and in a detailed treatment of the intermediate state.

Applying $\operatorname{curl} \mathbf{E} = -\dot{\mathbf{H}}/c$ to equation (B), one obtains

$$\operatorname{curl}[\mathbf{E} - (4\pi\lambda_L^2/c^2)\,\dot{\mathbf{J}}] = 0,$$

showing that $\qquad\qquad \mathbf{E} - (4\pi\lambda_L^2/c^2)\,\dot{\mathbf{J}} = \operatorname{grad}\phi,$

where $\phi$ is a scalar. In the most general case of a multiply connected superconductor or a superconducting portion of a current-carrying circuit, one cannot prove that $\phi$ vanishes. Hence (A) does not always follow from (B) and the perfect conductivity implicit in (A) and the perfect diamagnetism in (B) must be considered as independent postulates.

In a system of $N$ particles of charge $q$ described by the wave function

$$\Psi(r_1, r_2, \ldots, r_N),$$

the mean current density at a point $\mathbf{R}$ in the presence of a magnetic field

$$\mathbf{H}(\mathbf{r}_\alpha) = \operatorname{curl}\mathbf{A}(\mathbf{r}_\alpha) \qquad\qquad \text{(III.19)}$$

is given by

$$\mathbf{J}(\mathbf{R}) = \sum_{\alpha=1}^{N} \int_1 \cdots \int_N \left\{ \frac{q\hbar}{2im}[\Psi^*\nabla_\alpha\Psi - \Psi\nabla_\alpha\Psi^*] - \frac{q^2}{mc}\mathbf{A}(\mathbf{r}_\alpha)\,\Psi^*\Psi \right\} \delta(\mathbf{R}-\mathbf{r}_\alpha)\,dr_1\ldots dr_N. \quad \text{(III.20)}$$

In the absence of a field, $\mathbf{A}(\mathbf{r}_\alpha) \equiv 0$, $\Psi \equiv \Psi_0$, and the current density vanishes, so that

$$\sum_{\alpha=1}^{N} \int_1 \cdots \int_N \left\{ \frac{q\hbar}{2im}[\Psi_0^*\nabla_\alpha\Psi_0 - \Psi_0\nabla_\alpha\Psi_0^*]\,\delta(\mathbf{R}-\mathbf{r}_\alpha) \times \right.$$
$$\times\, dr_1\ldots dr_N \equiv 0. \qquad \text{(III.21)}$$

If, therefore, one assumes that the wave function $\Psi$ is perfectly rigid under the application of a magnetic field, that is, that $\Psi = \Psi_0$ always, then it follows that

$$\mathbf{J}(\mathbf{R}) = -\sum_{\alpha=1}^{N} \int_1 \cdots \int_N \frac{q^2}{mc}\mathbf{A}(\mathbf{r}_\alpha)\,\Psi^*\Psi\,\delta(\mathbf{R}-\mathbf{r}_\alpha)\,dr_1\ldots dr_N. \quad \text{(III.22)}$$

By defining a particle density

$$n(\mathbf{R}) \equiv \sum_{\alpha=1}^{N} \int_{1} \ldots \int_{N} \Psi^* \, \Psi \, \delta(\mathbf{R} - \mathbf{r}_\alpha) \, dr_1 \ldots dr_N, \quad \text{(III.23)}$$

equation III.22 can be written as

$$\mathbf{J}(\mathbf{R}) = -n(\mathbf{R}) \frac{q^2}{mc} \mathbf{A}(\mathbf{R}). \quad \text{(III.24)}$$

But if the particle density $n(\mathbf{R})$ is a sufficiently smooth function so that one can replace it by a constant $n$, then in view of the defining equation III.13, III.24 is seen to be identical to (B′).

Thus the London equation (B) or (B′) implies that the magnetic properties of a superconductor are due to a complete rigidity of the wave functions of the superconducting carriers. In F. London's own words ([2], p. 150): '... superconductivity would result if the eigenfunctions of a fraction of the electrons were not disturbed at all when the system is brought into a magnetic field $(H < H_c)$.'

A possible explanation of this is contained in the London equations themselves. The mean local value of the carriers' momentum in the presence of a field is given by

$$\mathbf{p} = m\mathbf{v} + (q/c)\mathbf{A},$$

which can be rewritten as

$$\mathbf{p} = (q/c)[(4\pi\lambda_L^2/c)\mathbf{J} + \mathbf{A}]. \quad \text{(III.25)}$$

In the same gauge as that leading to (B′), III.25 for a simply connected superconductor reduces to

$$\mathbf{p}_s = 0. \quad \text{(III.26)}$$

The London equation thus implies that superconductivity is due to a condensation of a number of carriers into a lowest momentum state $\mathbf{p}_s = 0$. By the uncertainty principle this requires the essentially unlimited spatial extension of the appropriate wave functions, and makes it impossible for them to be affected by local field variations.

It also follows from III.26 that

$$\mathbf{v}_s = -(q/mc)\mathbf{A}, \quad \text{(III.27)}$$

showing that in a simply connected superconductor the charge flow is entirely determined by the externally applied field, and exists only in its presence.

### 3.6. Quantized flux

F. London already observed ([2], p. 151] that the unlimited extension of the wave function of the superconducting charge carriers has a very fundamental consequence in a multiply connected superconductor. Consider, for example, a superconductor containing a hole. The wave functions must then be single valued along any closed path enclosing the hole. By analogy to the electronic wave functions in an atomic orbit one can then apply the Bohr-Sommerfeld quantization rules and require that for the superconducting charge carriers

$$\oint \mathbf{p} \cdot d\mathbf{l} = nh, \tag{III.28}$$

along any path enclosing the hole. According to III.25 this then means that

$$\oint \frac{4\pi\lambda^2}{c} \mathbf{J} \cdot d\mathbf{l} + \oint \mathbf{A} \cdot d\mathbf{l} = n\frac{hc}{q} \tag{III.29}$$

Since $\mathbf{H} = \mathrm{curl}\,\mathbf{A}$, the contour integral of $\mathbf{A}$ is equal to the surface integral of $\mathbf{H}$ over the area enclosed by the contour, and this in turn equals the magnetic flux $\Phi$ threading the contour:

$$\oint \mathbf{A} \cdot d\mathbf{l} = \int \int \mathbf{H} \cdot d\mathbf{S} = \Phi. \tag{III.30}$$

Thus

$$\oint \frac{4\pi\lambda^2}{c} \mathbf{J} \cdot d\mathbf{l} + \Phi = n\frac{hc}{q}. \tag{III.31}$$

London called the left-hand side of this equation a *fluxoid*, and we see that according to III.31 such a fluxoid is quantized in integral multiples of

$$\phi_0 \equiv \frac{hc}{q}. \tag{III.32}$$

Note that if the contour is taken at a distance from the hole large compared to the penetration depth $\lambda$, the current density vanishes, and the fluxoid is just equal to the total flux associated with the hole. This flux is thus seen to be quantized.

The quantization of flux was verified experimentally by Doll and Näbauer (1961) and by Deaver and Fairbank (1961). These experiments have shown that the quantum of flux is given by

$$\phi_0 = \frac{hc}{2e} \sim 2 \times 10^{-7} \text{ gauss-cm}^2.$$

This shows that $q = 2e$, that is, that the superconducting charge carriers are *pairs of electrons*. It has already been mentioned that indeed this is the fundamental premise of the microscopic theory.

A number of authors (Byers and Yang, 1961; Onsager, 1961; Bardeen, 1961b; Keller and Zumino, 1961; Brenig, 1961) have extended the London argument for flux quantization in a rigorous fashion. In particular Byers and Yang as well as Brenig have shown explicitly that the quantization is due to a periodicity of the free energy of the superconductor as a function of flux. The free energy of the normal phase is essentially independent of flux, and there must therefore occur a corresponding periodic variation of the critical temperature at which the free energies of the two phases are equal. This flux periodicity of $T_c$ has been observed by Little and Parks (1962) and by Groff and Parks (1966).

The fundamental importance of flux quantization in superconductors will be stressed in a number of subsequent chapters.

# CHAPTER IV

# The Pippard Non-local Theory

## 4.1. The penetration depth $\lambda$

The London equations lead to an exponential penetration of an externally applied magnetic field into a superconductor, so that the penetration can be characterized by the depth $\lambda$ at which the field has fallen to $1/e$ of its value at the surface. Quite in general, and independently of any particular set of electromagnetic equations for the superconductor, one can define the penetration depth for an infinitely thick specimen by

$$\lambda \equiv \frac{1}{H_e} \int\limits_0^\infty H(x)\, dx. \qquad (IV.1)$$

This would apply equally well to an exponentially decaying field as to one, improbable though it may be, which remains constant to a certain depth and then vanishes suddenly.

Shoenberg ([1], p. 140) has pointed out that in this way one can treat problems involving either very thick specimens (thickness $a \gg \lambda$) or very thin ones ($a \ll \lambda$) independently of a detailed knowledge of the appropriate electromagnetic equations. Using IV.1 to calculate the ratio of the magnetic susceptibility $\chi$ of a sample into which the applied field has penetrated, to the susceptibility $\chi_0$ of an identical sample from which the field is entirely excluded, he finds equations of which the following are applicable to a plate of thickness $2a$ in a uniform field parallel to its surface:

$$\chi/\chi_0 = 1 - \lambda/a \quad \text{for } a \gg \lambda, \qquad (IV.2)$$

$$\chi/\chi_0 = \alpha a^2/\lambda^2 \quad \text{for } a \ll \lambda. \qquad (IV.3)$$

The detailed form of the field penetration does not enter at all into IV.2 and all other equations for large specimens of other shapes, and does so only through the numerical parameter $\alpha$ in the equations for small specimens. As a consequence it is impossible to test the validity of any particular penetration law, such as, for example, the London relation III.17, by measurements on large specimens; with very small specimens this can only be done if one can determine absolute values of $a$ and of $\lambda$, which is very difficult. On the other hand, one can measure the variation of the susceptibility of large or of small specimens with any parameter affecting only $\lambda$: temperature, external field, impurity content, etc. One can then deduce directly the variation of $\lambda$ with the parameter in question, without having to make any assumptions about the true penetration law. The results of such measurements can therefore help to choose between different electromagnetic theories if these predict different parametric dependences of the penetration depth.

The oldest method of measuring the penetration depth consists of determining the magnetic susceptibility of samples with large surface to volume ratios to make the penetration effects appreciable. Shoenberg (1940) measured the temperature dependence of the susceptibility of a mercury colloid containing particles of diameter between 100 and 1000 Å. Désirant and Shoenberg (1948) used composite specimens consisting of about 100 thin mercury wires of diameter about $10^{-3}$ cm, and Lock (1951) carried out extensive measurements of the magnetic behaviour of thin films of tin, indium, and lead. Casimir (1940) suggested a method using macroscopic specimens in which the mutual inductance was measured between two coils closely wound around a cylinder of superconducting material. It was applied successfully by Laurmann and Shoenberg (1947, 1949), by Shalnikov and Sharvin (1948), and most recently with certain refinements by Schawlow and Devlin (1959), and by Maxfield and McLean (1965).

A superconductor has finite surface impedance at high frequency, and this impedance is limited in the superconducting phase by the penetration depth $\lambda$, as in the normal phase it is limited by the skin depth $\delta$. Pippard (1947a) was the first to use this as a means of measuring $\lambda$, and he and his collaborators have carried out a large number of experiments at different frequencies and varying experimental conditions (see Pippard, 1960). Basically all these measurements involve observing the change in the resonant frequency of a cavity containing the specimen when the specimen passes from the normal to the superconducting phase. At $T \ll T_c$, where $\lambda \ll \delta$, these changes are proportional to $\delta - \lambda$. If $\delta$ is independent of temperature, as is the case for a metal in the residual resistivity range, then any temperature variation of the observed changes must be due to the temperature variation of $\lambda$. Dresselhaus *et al.* (1964) use instead of a cavity a rutile resonator to which the specimen is coupled, and also observe changes in the resonant frequency.

Schawlow (1958), Jäggi and Sommerhalder (1959, 1960), and most recently Erlbach *et al.* (1960) have measured the penetration of a magnetic field through a thin cylindrical film of thickness less than the penetration depth.

### 4.2. The dependence of $\lambda$ on temperature and field

According to the London theory, an external magnetic field penetrates into a superconductor to a depth characterized by (see equation III.13)

$$\lambda_L = (mc^2/4\pi n_s e^2)^{1/2},$$

where $n_s$ is the number density of the superconducting electrons. It is reasonable to expect this to be the only temperature-dependent factor in this defining equation, and in fact the Gorter-Casimir two-fluid model assumes that

$$n_s(t) = \mathscr{W}(t) n_s(0), \tag{IV.4}$$

where $\mathscr{W}(t)$ is the order parameter.

The temperature dependence of $\mathscr{W}$ is given by II.25′, so that substituting this and IV.4 into the defining equation for the penetration depth one obtains

$$\lambda_L(t) = \lambda_L(0)/(1 - t^4)^{1/2}, \tag{IV.5}$$

where
$$\lambda_L(0) = (mc^2/4\pi n_s(0) e^2)^{1/2} \tag{IV.6}$$

is the penetration depth at $T = 0°K$. Very near $T_c$, IV.5 can be written as

$$\lambda_L(t) = \frac{\lambda_L(0)}{2}(1 - t)^{-1/2}. \tag{IV.7}$$

Daunt *et al.* (1948) were the first to point out that the empirical temperature variation of the penetration depth can indeed be represented to a very high degree of approximation by IV.5. This is a striking success of the phenomenological theories discussed in the preceding chapters. Close inspection of the recent very precise measurements, however, shows a small deviation from IV.5 at $t < 0.8$, which becomes particularly pronounced at low temperatures. This deviation is barely discernible in the normal plot of $\lambda(t)$ vs. $y(t)$, where $y(t) \equiv (1 - t^4)^{-1/2}$, but is displayed strikingly in Figure 13, which shows for Schawlow's results (1958) the variation of the slope,

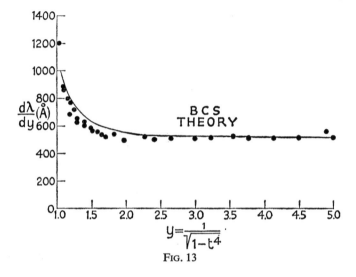

Fig. 13

$d\lambda/dy$, with $y(t)$. The solid line indicates the values of $d\lambda/dy$ calculated by Miller (1960) on the basis of the BCS theory; the experimental results appear to deviate *less* from IV.5 than is predicted by theory. Furthermore it appears that in impure specimens no deviation from IV.5 can be found at all (Waldram, 1964).

The slope of $\lambda(t)$ plotted as a function of $y(t)$, as well as the intercept, yield values for $\lambda(0)$ if one ignores the small deviations from IV.5. Appropriate empirical values for pure bulk samples are shown in Table II. They exceed by a factor of about five what one would expect from the London definition IV.6, unless one makes rather unlikely assumptions of low densities of superconducting electrons or of a large effective mass. Experiments on very small samples, and measurements on impure metals, yield even higher values of $\lambda(0)$, although none of the factors in IV.6 appear to depend on size or purity. This failure of the London theory will be discussed in the following sections of this chapter.

TABLE II

| Element | $\lambda(0)$ (Å) | Reference |
|---------|------------------|-----------|
| Al | 500 | Faber and Pippard, 1955a |
| Cd | 1300 | Khaikin, 1958 |
| Hg | 380–450* | Laurmann and Shoenberg, 1949 |
| In | 640 | Lock, 1951 |
| Nb | 470 | Maxfield and McLean, 1965 |
| Pb | 390 | Lock, 1951 |
| Sn | 510 | Pippard, 1947; Laurmann and Shoenberg, 1949; Lock, 1951 |
|  | 470–600* | Schawlow and Devlin, 1959 |
| Tl | 920 | Zavaritskii, 1952 |

* Anisotropy.

Pippard (1950) investigated the change of the penetration of a small, r.f. field (9·4 kMc/s) at a given temperature as an external d.c. field $H_e$ is raised from zero to the critical value. This change, divided by the penetration depth in zero field, is plotted against temperature in Figure 14. There are clearly two effects: one at low temperatures (which Bardeen (1952, 1954) has shown to follow from an extension of the London equations to include non-linear terms), and one near $T_c$. This latter involves a change in $\lambda$ with $H$ in just that region in which $\lambda$ varies appreciably with temperature. By a thermodynamic derivation Pippard has shown that this temperature variation of $\lambda$ leads to a dependence of the superconducting entropy on field. He finds that

$$S(H_e) - S(0) = \frac{\lambda(0) \, A H_e^2}{4\pi T_c} [t^3/(1 - t^4)] \qquad (IV.8)$$

for a superconductor of total surface $A$, assuming IV.5 to hold. Near $T_c$ this change contributes as much as one-fourth of the total entropy

4

difference between the normal and superconducting phases, which is quite considerable.

Pippard pointed out that to assume that the entire entropy change takes place in the thin layer into which the field penetrates would, therefore, result in an unreasonably high entropy density in this layer. Yet this is just what one is led to believe by the London model, according to which the superconducting wave functions or, in two-fluid language, the corresponding order parameter $\mathscr{W}$, remains rigidly unchanged by the application of an external field. Any change in the

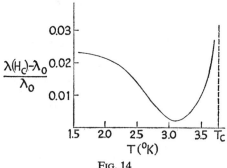

Fig. 14

thermodynamic functions with field must therefore be confined to the thin layer into which the field penetrates.

Recent measurements of the field dependence of the penetration depth at 1 and 3 kMc/s (Spiewak, 1959; Richards, 1960, 1962; Pippard, 1960; Dresselhaus *et al.*, 1964) have shown that this effect has certain unexpectedly complicated features. Not only is the magnitude of the change frequency dependent, but even its sign can change under certain conditions. In particular there can be an increase in the penetration depth when the applied field is parallel to the specimen surface, and at the same frequency a decrease when the field is perpendicular. Bardeen (1958) and Pippard (1960) have suggested that these complexities may be due to field induced deviations of the superconducting and normal electron densities from their equilibrium values. Maki (1964c), however, indicates that most of the observed effects follow directly from the microscopic theory.

## 4.3. The range of coherence

The unreasonably high entropy density in the surface layer led Pippard (1950) to propose a basic modification of the London model, according to which the order parameter changes gradually over a certain length $\xi$, which he calls the range of coherence of the superconducting wave functions. In terms of the microscopic theory this distance can be considered as the typical size of the Cooper pairs. Any change in the thermodynamic functions of course extends over as wide a region as the change in the order parameter, and thus a value $\xi \gg \lambda$ would correspond to a more reasonably small value of the field-induced entropy density.

Pippard (1950) obtained an estimate of the range of coherence of the order parameter by minimizing the Gibbs free energy of the superconductor in the presence of an external field. The resulting relation between the fractional change of the penetration depth and the ratio $\lambda(0)/\xi$ allows him to estimate from his experimental data on the field effect on $\lambda$ that $\xi \approx 20\lambda(0) \approx 10^{-4}$ cm. Such a distance is much larger than the smallest colloidal specimen or the thinnest films in which superconductivity is still known to exist, and it is therefore of great importance to realize that, to quote Pippard (1950, p. 220): 'the range of order must therefore not be regarded as a minimum range necessary for the setting up of an ordered state, but rather as the range to which order will extend in the bulk material'.

Strong support is given to the existence of this range of coherence by the extreme sharpness of the superconducting transition under suitable conditions. De Haas and Voogd (1931) have observed resistive transitions in single crystals of tin taking place within a range of one millidegree, and a sharpness approaching this value has come to be the criterion for the quality of a specimen of suitable shape and orientation. Applying a simple statistical argument, Pippard shows that fluctuations would create a broader transition unless the superconductivity of a bulk sample can be created or destroyed only over an entire domain of diameter $\approx 10\lambda(0)$.

In the next section as well as in 6.5 it will be shown that the range of coherence is much smaller than $10^{-4}$ cm in low mean free path alloys as well as in certain pure metals. For such materials one would expect

a broadened transition even for ideally homogeneous samples.
Goodman (1962c) has discussed this in some detail.

### 4.4. The Pippard non-local relations

In 1953 Pippard measured the penetration depth in a series of dilute
alloys of indium in tin, and found that the decrease in the normal
electronic mean free path of the metal was accompanied by an

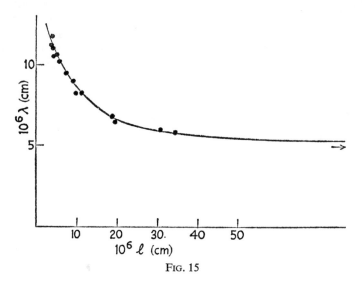

Fig. 15

appreciable rise in the value of $\lambda(0)$. This has been confirmed by
Chambers (1956), and by Waldram (1964), whose results are shown
in Figure 15. Such a dependence of $\lambda(0)$ on the mean free path is quite
incompatible with the London model, for clearly none of the parameters in the defining equation IV.6 varies appreciably with the
electronic mean free path.

This experimental result, added to the previous questions which had
been raised about the correctness of the London phenomenological
treatment, led Pippard (1953) to develop a fundamental modification
of this model, based on the concept of the range of coherence of the

superconducting phase. The basic London equations, it will be remembered, lead to the relation

$$\mathbf{J(R)} = -\frac{c}{4\pi\lambda_L^2}\mathbf{A(R)}, \qquad \text{(IV.10)}$$

where $\lambda_L = mc^2/4\pi ne^2$, so that one can also write this as

$$\mathbf{J(R)} = -\frac{ne^2}{m}\mathbf{A(R)}. \qquad \text{(IV.11)}$$

One way to introduce a dependence of the penetration depth on the electronic mean free path is to write

$$\mathbf{J(R)} = -\frac{ne^2}{m}\frac{\xi(l)}{\xi_0}\mathbf{A(R)}, \qquad \text{(IV.P1)}$$

where $\xi_0$ is a constant of the superconductor in question, and $\xi(l)$ a parameter depending on the mean free path $l$. It is evident from the analysis in Chapter III that IV.P1 leads to an expression for the field penetration into a semi-infinite slab which has the London form:

$$H(x) = H_e \exp(-x/\lambda),$$

but where now
$$\lambda = \lambda_L\bigg/\sqrt{\left(\frac{\xi_0}{\xi(l)}\right)}. \qquad \text{(IV.12)}$$

As experimentally $\lambda$ is found to increase with decreasing $l$, it is clear that $\xi(l)$ must decrease as $l$ decreases.

As a first step toward a modification of the London model, Pippard identifies $\xi_0$ with the range of coherence of the pure superconductor, and assumes that $\xi(l)$ tends toward this value as $l \to \infty$, but that $\xi(l) \to l$ as $l \to 0$. This is the case if, for example,

$$\frac{1}{\xi(l)} = \frac{1}{\xi_0} + \frac{1}{\alpha l}, \qquad \text{(IV.13)}$$

where $\alpha$ is a constant of order unity. $\xi(l)$ is thus an effective range of coherence which has a size ($\approx 10^{-4}$ cm) characteristic of the metal in a pure superconductor, but which becomes limited by the normal electronic mean free path as the latter becomes much smaller than $10^{-4}$ cm.

*Superconductivity*

These equations satisfactorily explain the onset of a mean free path effect on the penetration depth at a critical value of $l$, as found by Pippard and others, as well as the very large penetration depth values obtained from experiments where $l$ is limited by boundary scattering. They do not, however, satisfactorily explain the finding that $\lambda(0)$ in pure, bulk superconductors exceeds the London value IV.6 by a factor of four to five. According to Pippard this is because IV.P1 does not correctly describe the relation between current density and the vector potential in such a case. P1 still implies, as does equation III.B′, the basic London idea of a wave function which is completely rigid under the application of an external field because the electronic momenta are ordered or correlated over an infinite distance. Thus the distance over which H and A vary is quite immaterial; the same kind of relation would hold if the field varied very slowly as if it varied very rapidly. But according to Pippard the range of momentum coherence is not infinite but only about $10^{-4}$ cm, so that the electromagnetic response of the superconductor should be affected profoundly if the field varies rapidly over this distance. A relation like P1 could apply only if the field varied slowly over a distance of the order of $\xi$.

The situation is somewhat analogous to the problem of electrical conductivity in a normal conductor, for which the relation

$$\mathbf{J(R)} = \sigma(l)\,\mathbf{E(R)} \tag{IV.14}$$

is valid only if $\mathbf{E(R)}$ varies slowly over a distance of the order of $l$. An applied alternating field penetrates only a finite distance, $\delta$, which varies inversely as the square root of the frequency. At sufficiently low temperatures and high frequencies, the electronic mean free path in the normal metal may be longer than this skin depth, so that electrons may spend only part of the time between collisions in the field penetrated region. Pippard (1947a) showed how this makes the electrons less effective as carriers of current and leads to a higher surface resistance, as observed by H. London (1940) and Chambers (1952). Under these conditions Ohm's law (IV.14) can no longer be a valid approximation; the current at a point must be determined by the integrated effect of the field over distances of the order of the mean free path (see Pippard, 1954). The details of this so-called anomalous skin effect were worked out by Reuter and Sondheimer (1948), who derived that

$$\mathbf{J(R)} = \frac{3\sigma}{4\pi l} \int \frac{\mathbf{R(R \cdot E)}\,e^{-R/l}\,d\tau}{R^4}, \qquad \text{(IV.15)}$$

where $\sigma$ is the d.c. conductivity and $l$ the mean free path. The form of this equation ensures that in the case of a rapidly varying field the current density at a point $\mathbf{R}$ is determined by the integral of the field over a distance comparable to the mean free path $l$.

In a superconductor of range of coherence $\xi$, the current density at a point in the case of a rapidly varying field should also be determined by an integral of the field over a distance of the order of $\xi$, and not, as is implicit in the London equation as well as in P1, by the field variation over a quasi-infinite distance. Because of this analogy to the anomalous conduction in a normal metal, and because some special solutions of equation IV.15 were already known, Pippard (1953) proposed as the basic relation for the electromagnetic response of a pure superconductor the equation

$$\mathbf{J(R)} = -\frac{3ne^2}{4\pi\xi_0 m} \int \frac{\mathbf{R(R \cdot A)}\,e^{-R/\xi}\,d\tau}{R^4}. \qquad \text{(IV.P2)}$$

Somewhat misleadingly, as this erroneously implies that the basic London equation is a truly local relation, P2 is called the Pippard non-local relation.

This relation leads to a reversal of the phase of the magnetic field penetrating into a superconductor (Pippard, 1953). Drangeid and Sommerhalder (1962) have observed this effect.

The validity of P2 is strongly supported by Bardeen's proof ([5], pp. 303 ff.) that an energy gap in the single electron spectrum requires a non-local relation between current density and vector potential. In fact the BCS theory leads to a relation entirely equivalent to P2 if one assumes

$$\xi_0 = \hbar v_0 / \pi\epsilon(0), \qquad \text{(IV.16)}$$

where $2\epsilon(0)$ is the energy gap at $0°$K. Substituting the BCS value $2\epsilon(0) = 3{\cdot}52 k_B T_c$, IV.16 becomes

$$\xi_0 = 0{\cdot}18\,\hbar v_0 / k_B T_c. \qquad \text{(IV.17)}$$

This is just the expression IV.9 for the range of coherence derived from an uncertainty principle argument, with $a = 0.18$.

From P2, the penetration depth $\lambda$ as defined by IV.1 can be evaluated explicity in two limiting cases:

$$\lambda = \sqrt{(\xi_0/\xi)}\lambda_L \qquad \text{for } \xi \ll \lambda, \text{ (London limit),} \quad \text{(IV.18a)}$$

$$\lambda_\infty = \left[\frac{\sqrt{3}}{2\pi}\xi_0\lambda_L^2\right]^{1/3} \qquad \text{for } \xi \gg \lambda, \text{ (Pippard limit).} \quad \text{(IV.18b)}$$

The second of these is the one applicable to the case of an infinite mean free path, and correctly predicts a penetration depth into very pure superconductors which is much larger than the London value. IV.18a is identical to IV.12 obtained directly from P1. This is of course a reflection of the fact that P1 is the limiting form for $\xi \ll \lambda$ of the more general equation P2.

Equations IV.18 show that the range of coherence of a superconductor can be calculated from absolute values of the penetration depth. Faber and Pippard (1955a) have in this way obtained values of $2.1 \times 10^{-5}$ cm for tin, and $12.3 \times 10^{-5}$ cm for aluminium. These values differ very much, but when substituted into equation IV.9 together with known values of $T_c$ and $v_0$ (from anomalous skin effect data [Chambers, 1952]), both correspond to $a = 0.15$. This is in striking agreement with the BCS value of $0.18$, as cited in IV.17. A later chapter will mention how measurements of the transmission of infrared radiation through thin superconducting films lends further strong support to this value.

Peter (1959) has solved the Pippard non-local relation P2 for the case of cylindrical superconducting films of thickness $d < \lambda$ and radius $r$. He finds that an external field $H_e$ penetrates through the film to a value $H_i$ such that

$$H_e/H_i = [3rd^2/8\lambda^2\xi_0]F(\xi/d). \quad \text{(IV.19)}$$

$\xi_0$ is the range of coherence in a specimen of unlimited mean free path, and can be calculated from IV.17; $\xi$ is the actual range of coherence in the film, and $\lambda$ should be the London penetration depth as calculated from IV.5 and IV.6. Schawlow (1958), however, has shown that good agreement with his measurement on tin films can be found by substituting for $\lambda$ the empirical value for bulk samples (510 Å) and considering $\xi$ as being determined by the size-limited mean free path of the electrons in the films. A similar analysis has been used by Sommerhalder (1960).

It is now generally accepted that whenever one applies the equation of the Pippard theory (or those of the Ginzburg-Landau treatment to be discussed presently) to the case of small or impure specimens, one obtains good agreement by using for the ideal penetration depth in a bulk sample, not the London value $\lambda_L$ but rather the depth determined experimentally. For example, the results of Whitehead (1956) on the magnetic properties of mercury colloids were shown by Tinkham (1958) to be in excellent agreement with the prediction of the London limit of the Pippard theory if one modifies equation IV.18b and writes

$$\lambda = \lambda_b \bigg/ \sqrt{\left(\frac{\xi_0}{\xi(l)}\right)}, \qquad \text{(IV.20)}$$

where $\lambda_b$ is now the empirical penetration depth for a bulk sample and takes the place of the London value $\lambda_L$. The mean free path $l$ is limited by boundary as well as by impurity scattering. Ittner (1960a) has similarly found that such a modification of the Pippard equations adequately predicts the results of the observations by Blumberg (1962) of the critical field of moderately thin films. In analysing the magnetic behaviour of small (or very impure) specimens, for which $\xi \approx l \ll \lambda$, it is thus in general possible to obtain adequate precision without attempting to solve the difficult relation IV.P2. Instead one can use IV.20 to calculate the penetration depth, and then substitute this value of $\lambda$ into the London equation IV.10.

In discussing the mean free path dependence of the coherence length one must remember that it is related to the behaviour of a superconductor in two subtly different ways. One of these, as mentioned in Section 4.3, is the distance over which the order parameter of the superconducting phase varies. It is this aspect which, for example, in Chapter VI will enter into the discussions of the width of a boundary between the normal and superconducting phases. It follows from Gor'kov's analysis of the influence of impurities

(Gor'kov 1959b) that the mean free path dependence of this aspect of the range of coherence is given by

$$\xi = \xi_0 \chi^{-1/2}(l) \qquad \text{(IV.21)}$$

$\chi(l)$ is a function of the mean free path shown graphically by Gor'kov and approximated to within about 20 per cent by the simple expression (Douglass and Falicov, 1964)

$$\chi(l) \sim \left( \frac{1}{\xi_0} + \frac{1}{l} \right). \qquad \text{(IV.22)}$$

In the limit $l \ll \xi_0$, IV.21 thus reduces to

$$\xi = \sqrt{(\xi_0 l)}. \qquad \text{(IV.23)}$$

The relatively slow variation of this aspect of the coherence length with mean free path is essentially due to the fact that not every electronic collision destroys the superconducting coherence.

The other aspect of the range of coherence is that it determines the distance over which the magnetic field or the vector potential at a given point influences the current density. This is expressed by the Pippard equation IV.P2. What is important in this application is the actual mean distance between electron collision, so that now equation IV.13 applies. This means that

$$\xi \approx l \qquad \text{(IV.24)}$$

for $l \ll \xi_0$. It is this mean free path dependence which enters, for example, into equation IV.20.

# CHAPTER V

# The Ginzburg-Landau
# Phenomenological Theory

## 5.1. Basic formulation

In 1950 Ginzburg and Landau (G-L) introduced a phenomenological approach to superconductivity which, like that of Pippard, modifies the absolute rigidity of the superconducting order parameter or wave function which is implicit in the London model. Although the theory was originally formulated so as to reduce always to the 'local' London equations in zero field, Bardeen (1954) has shown that it can be modified so as to be compatible with a non-local equation of the Pippard type. Furthermore, Gor'kov (1959, 1960) has derived the G-L equations, under certain conditions, from his formulation of the BCS theory.

G-L introduce an order parameter $\psi$ which they normalize so as to make $|\psi|^2 = n_s$, where $n_s$ is the density of the superconducting electrons. $\psi$ is thus a kind of 'effective' wave function of the superconducting electrons. According to the general Landau-Lifshitz theory of phase transitions (1958), the free energy of the superconductor depends only on $|\psi|^2$ and can be expanded in series form for temperatures near $T_c$. In the absence of an external field, the superconducting free energy (per unit volume, as are all equations listed) is then

$$G_s(0) = G_n(0) + \alpha|\psi|^2 + (\beta/2)|\psi|^4. \qquad (V.1)$$

Minimizing the free energy with respect to $|\psi|^2$ yields the zero field equilibrium value

$$|\psi_0|^2 = -\alpha/\beta, \qquad (V.2)$$

from which $\qquad G_s(0) - G_n(0) = -\alpha^2/2\beta. \qquad (V.3)$

In the immediate vicinity of $T_c$ one can assume that the coefficients $\alpha$ and $\beta$ have the simple form

$$\left.\begin{array}{l} \alpha(T) = (T_c - T)(\partial\alpha/\partial T)_{T=T_c} \\[2mm] \beta(T) = \beta(T_c) \equiv \beta_c \end{array}\right\} \quad (V.4)$$

and

With these one then finds from V.3, remembering that the free energy difference between the phases equals the magnetic energy, that

$$H_c^2 = \frac{4\pi\alpha^2}{\beta} = \frac{4\pi(T_c - T)^2}{\beta_c}\left(\frac{\partial\alpha}{\partial T}\right)_{T=T_c}^2. \tag{V.5}$$

Near $T_c$, $H_c$ indeed is known to vary linearly with $(T_c - T)$, so that the correctness of equation V.5 justifies the assumptions V.4. All further thermodynamic manipulations are now possible, but they and all other conclusions drawn using V.4 are restricted to temperatures very near $T_c$. Both Bardeen (1954) and Ginzburg (1956a) have considered extensions of the model to the full superconducting range by introducing different forms for $\alpha(T)$ and $\beta(T)$, the former using expressions based on the Gorter–Casimir two-fluid model. These extensions, which do not follow from the microscopic theory, have since been superseded by the work to be described in Section 5.4.

The outstanding contribution of the G-L model in any temperature range arises from its ability to treat the superconductor in an external field $H_e \approx H_c$. The free energy $G_s(H_e)$ is now increased not only by the usual volume term $H_e^2/8\pi$, but also by an extra term connected with the appearance of a gradient of $\psi$, as $\psi$ is not completely rigid in the presence of $H_e$. Such a gradient would contribute to the energy in analogy to the kinetic energy density in quantum mechanics which depends on the square of the gradient of the wave function. Introducing this extra energy is equivalent to requiring that $\psi$ not change too abruptly. One is thus led to a concept of gradual, extended variations of the superconducting order parameter quite analogous to Pippard's model of the range of coherence.

In order to preserve gauge-invariance, G-L assume the extra energy term to be

$$\frac{1}{2m}\left[-i\hbar\nabla\psi - \frac{e^*}{c}\mathbf{A}\psi\right]^2, \tag{V.6}$$

where $\mathbf{A}$ is the vector potential of the applied field, and $e^*$ a charge which, as stated in the original version of the theory, 'there is no reason to consider as different from the electronic charge'. Modifications of this view will be discussed presently.

G-L thus write

$$G_s(H_e) = G_s(0) + \frac{H_e^2}{8\pi} + \frac{1}{2m}\left[-i\hbar\nabla\psi - \frac{e^*}{c}\mathbf{A}\psi\right]^2. \qquad (V.7)$$

One must now minimize this with respect to both $\psi$ and to $\mathbf{A}$, which leads to the two equilibrium equations:

$$\frac{1}{2m}\left(-i\hbar\nabla - \frac{e^*}{c}\mathbf{A}\right)^2\psi + \frac{\partial G_s(0)}{\partial\psi^*} = 0, \qquad (V.G\text{-}L1)$$

$$\nabla^2\mathbf{A} = -\frac{4\pi}{c}\mathbf{J}_s = \frac{2\pi i e^*\hbar}{mc}[\psi^*\,\nabla\psi - \psi\nabla\psi^*] +$$

$$+\frac{4\pi e^{*2}}{mc^2}|\psi|^2\mathbf{A}. \qquad (V.G\text{-}L2)$$

In a very weak field, $H \approx 0$, the function $\psi$ remains practically constant (that is, rigid), $\nabla\psi = 0$, $\psi \approx \psi_0$, and G-L2 reduces to

$$\nabla^2\mathbf{A} \approx \frac{4\pi e^{*2}}{mc^2}|\psi_0|^2\mathbf{A} = \frac{4\pi e^{*2}}{mc^2}n_s\mathbf{A}. \qquad (V.8)$$

Here, as in V.2, the subscript 0 denotes the zero field value. This of course is just London's expression (B). Non-local versions of the G-L treatment are obtained by substituting an integral expression for the second term in V.6. In its present local form the G-L treatment is restricted to temperatures near $T_c$ for two reasons: in the first place because of the simple forms assumed for the functions $\alpha(T)$ and $\beta(T)$, and secondly because only near $T_c$ is $\lambda \gg \xi_0$, and can the non-local electromagnetic character of superconductivity be ignored.

The set G-L1 and G-L2 of coupled non-linear equations in $\psi$ and $\mathbf{A}$ have been solved for essentially one-dimensional problems. Taking the $z$-axis to be normal to the infinite superconducting boundary, the field **H** along the $y$-axis, and the current $\mathbf{J}_s$ and potential **A** along the $x$-axis, one obtains (using V.1)

$$\frac{d^2\psi}{dz^2} - \frac{2m}{\hbar}\alpha\left(1 + \frac{e^{*2}}{2mc^2\alpha}A^2\right)\psi - \frac{2m}{\hbar^2}\beta\psi^3 = 0, \qquad \text{(V.9)}$$

$$\frac{d^2A}{dz^2} - \frac{4\pi e^{*2}}{mc^2}\psi^2 A = 0. \qquad \text{(V.10)}$$

Note that with this geometry

$$\mathbf{H} = \operatorname{curl}\mathbf{A} = \frac{dA}{dz}.$$

The meaning of tnese equations becomes clearer by introducing a dimensionless parameter $\kappa$ defined by

$$\kappa^2 \equiv \frac{2e^{*2}}{\hbar^2 c^2}H_c^2\lambda_0^4, \qquad \text{(V.11)}$$

where

$$\lambda_0^2 \equiv \frac{mc^2}{4\pi e^{*2}\psi_0^2}. \qquad \text{(V.12)}$$

The subscript 0 again denotes zero field. $\kappa$, $\lambda_0$, and $H_c$ are the three parameters of the G-L theory which are to be determined experimentally, and in terms of which various field and size effects can be expressed†. $H_c$ is the bulk critical field. $\lambda_0$ is the empirical penetration depth of a superconductor in the weak field limit, and is the quantity which through equation V.12 determines the zero field equilibrium value of the order parameter $\psi_0^2$. For a bulk sample containing impurities $\lambda_0$ increases, as was discussed in the previous chapter, and this in turn affects both $\psi_0$ and $\kappa$.

$\kappa$ can be determined in a number of ways, two of which follow directly from the defining equation V.11. In the immediate vicinity of $T_c$, the experimental variation of $\lambda_0(t)$ can be expressed by IV.7:

$$\lambda_0(t) = \frac{\lambda_0(0)}{2}(1-t)^{-1/2} = \frac{\lambda_0(0)}{2}\left(\frac{T_c}{\Delta T}\right)^{1/2} \quad \text{where } \Delta T \equiv T_c - T.$$

† Since $e^* = 2e$, only two of the three parameters are independent.

Also one can write

$$H_c = \left| \frac{dH_c}{dT} \right|_{T=T_c} \times \Delta T,$$

so that

$$\kappa^2 = \frac{e^{*2}}{8\hbar^2 c^2} \left| \frac{dH_c}{dT} \right|_{T=T_c}^2 \times T_c^2 \times \lambda_0^4(0). \qquad (V.13)$$

Thus $\kappa$ is seen to be temperature independent, at least for $T \approx T_c$.

One can also use the expression for the penetration depth derived from the BCS theory to be, very near $T_c$:

$$\lambda_0(t) = \frac{\lambda_L(0)}{\sqrt{2}}(1-t)^{-1/2} = \frac{\lambda_L(0)}{\sqrt{2}}\left(\frac{T_c}{\Delta T}\right)^{1/2},$$

so that

$$\kappa^2 = \frac{e^{*2}}{4\hbar^2 c^2} \left| \frac{dH_c}{dT} \right|^2 \times T_c^2 \times \lambda_L^4(0), \qquad (V.14)$$

where now $\lambda_L(0)$ is the London penetration depth calculated from III.13 using the actual free electron density. In principle this can be obtained from the normal state anomalous skin resistance (Chambers, 1962), but this does not yield reliable results.

Another method of calculating $\kappa$ for a given superconductor is to use results on supercooling, as will be discussed in a subsequent section. Ginzburg (1955) pointed out already before the formulation of the BCS theory that values as calculated from V.13 and V.14 could be made to agree very well with those obtained from supercooling data by taking $e^* = 2$ or $3e$. More recently Gor'kov (1958) has formulated the electromagnetic equation of the BCS microscopic theory in terms of Green's functions, and was able to show (1959, 1960) that the G-L equations G-L1 and G-L2 are identical to his expressions near $T_c$ when $\psi$ is taken to be proportional to the energy gap, and when one takes $e^* = 2e$. This again is an indication that the current carriers in superconductivity are the doubly charged Cooper pairs. With this value of $e^*$, V.13 and V.14 yield

$$\kappa = 1 \cdot 08 \times 10^7 \left| \frac{dH_c}{dT} \right|_{T=T_c} T_c \lambda_0^2(0), \qquad (V.13')$$

and $$\kappa = 2 \cdot 16 \times 10^7 \left| \frac{dH_c}{dT} \right|_{T=T_c} T_c \lambda_L^2(0). \qquad \text{(V.14')}$$

For tin, the first of these yields $\kappa = 0 \cdot 158$, the second $0 \cdot 149$, two values which are in excellent agreement. For indium, however, the respective values are $0 \cdot 112$ and $0 \cdot 051$ (Davies, 1960; Faber, 1961). For aluminium, the equations yield $0 \cdot 05$ and $0 \cdot 01$ (Davies, 1960). This lack of agreement may be in part due to errors in anomalous skin-effect measurements used to evaluate $\lambda_L(0)$, and in part, particularly in the case of aluminium, due to the large value of $\xi_0$, because of which local conditions hold only over a very small range of temperatures near $T_c$. In general, this range is defined by the criterion $T - T_c \leq \kappa^2 T_c$. Hence the values of $\kappa$ calculated from supercooling are probably the most reliable, providing that the effects of the surface sheath (see Section 7.7) have been taken into account.

In terms of the parameters $\kappa$, $\lambda_0$, and $H_c$, equation (V.9) reduces to

$$\frac{d^2 \psi}{dz^2} = \frac{\kappa^2}{\lambda_0^2} \left\{ - \left( 1 - \frac{A^2}{2H_c^2 \lambda_0^2} \right) \psi + \frac{\psi^3}{\psi_0^2} \right\}. \qquad \text{(V.9')}$$

Far from the phase boundary, for $z \to \infty$, $\psi^2 = \psi_0^2$, and

$$\frac{d\psi}{dz} = A = H = 0.$$

At the boundary, $z = 0$, V.9' is satisfied in the absence of an external field ($A \equiv 0$) by $\psi^2 = \psi_0^2$; $d\psi/dz = 0$. In other words, the presence of the phase boundary as such has no influence on the function $\psi$, which has the same value $\psi_0$ everywhere. In the presence of an external field $H_e$, however, this solution no longer applies, and one must integrate V.9' and V.10 with the boundary condition $\psi^2 = \psi_0^2$ for $z \to \infty$, and the condition $H = dA/dz = H_e$, and $d\psi/dz = 0$ for $z = 0$. This integration cannot be carried out exactly. Neglecting higher order terms, however, one finds equations for $\psi$ and for $A$ as functions of $z$. At $z = 0$, the value of $\psi$ is

$$\frac{\psi(H_e)}{\psi_0} = 1 - \frac{\kappa}{4} \frac{H_e^2/H_c^2}{(\kappa + \sqrt{2})}. \qquad \text{(V.15)}$$

With values of $\kappa \approx 0.1$, this equation predicts a decrease of $\psi$ by only about 2–3 per cent when $H_e = H_c$. It is not surprising, therefore, that the change in penetration depth with field is also very small. This can be calculated formally by using the defining equation IV.1 from which one finds that, with a weak measuring field normal to $H_e$:

$$\lambda(H_e) = \lambda_0 \left\{ 1 + \frac{\kappa(\kappa + 2\sqrt{2})}{8(\kappa + \sqrt{2})^2} \frac{H_e^2}{H_c^2} \right\}$$

$$\approx \lambda_0 \left\{ 1 + \frac{\kappa}{4\sqrt{2}} \frac{H_e^2}{H_c^2} \right\} \text{ for small } \kappa. \qquad \text{(V.16)}$$

For a measuring field parallel to $H_e$, the effect is tripled.

It is evident that in the limit $\kappa \to 0$, the effect of the external field on $\psi$ and on $\lambda$ vanishes, so that one returns to a situation formally equivalent to the London picture. It must however be noted that even for $\kappa = 0$, $\psi_0^2$ is deduced from the empirical value of $\lambda_0$. As a result one can in certain cases allow $\kappa$ to vanish without necessarily reducing the G-L treatment to the London one. Examples of this will be shown in the following sections.

### 5.2. Size effects on the critical field

When one of the dimensions of a superconducting specimen becomes comparable to the penetration depth, its critical magnetic field becomes much higher than that of a bulk sample of the same material at the same temperature. This follows already from the basic Gorter-Casimir thermodynamic description, according to which the free energy difference per unit volume between the superconducting and normal phases is

$$G_n(0) - G_s(0) = \frac{H_c^2}{8\pi}. \qquad \text{(V.17)}$$

In an external field $H_e$ a superconductor acquires an effective magnetization $M(H_e)$ and becomes normal when

$$\int_0^{H_e} M(H_e) \, dH_e = \frac{H_c^2}{8\pi}. \qquad \text{(V.18)}$$

The integral is the area under the magnetization curve, and it was pointed out in Chapter III that for any ellipsoidal specimens V.18

5

was satisfied when $H_e = H_c$. Actually this is true only when one neglects the penetration of the external field into the sample, which lowers the effective magnetization and the susceptibility of the sample, as shown by equations IV.2 and IV.3. The susceptibility determines the initial slope of the magnetization curve; a lower $\chi$ means that the curve has to go to a higher critical field to satisfy equation V.18. Clearly, assuming this curve to remain linear with slope $\chi$ right up to a critical field $H_s$:

$$\frac{H_s^2}{H_c^2} = \frac{\chi_0}{\chi}, \tag{V.19}$$

and

$$\frac{H_s}{H_c} = 1 + \frac{\lambda_0}{2a} \quad \text{for } a \gg \lambda_0, \tag{V.20}$$

$$\frac{H_s}{H_c} = \sqrt{3}\frac{\lambda_0}{a} \quad \text{for } a \ll \lambda_0, \tag{V.21}$$

using the London equation to evaluate $\alpha = \frac{1}{3}$ in IV.3 (Ginzburg, 1945; [1], p. 172). Similar expressions can be derived for spherical and cylindrical samples. The resulting equations agree well with the frequently observed enhancement of the critical field in small specimens when one uses for the penetration depth $\lambda_0$ the appropriate Pippard value as calculated from IV.20. This is a good example of how expressions derived from the London model can be used with the modified value of $\lambda$ (Tinkham, 1958).

The field enhancement calculated from the Ginzburg-Landau theory leads to nearly identical results. The essential difference is that because of the additional terms V.6 in the G-L free energy of the superconductor, the penetration depth increases in the presence of an external field (see equation V.16), so that the critical field for small samples becomes even higher. For thin films of thickness $2a$ the critical field is

$$\frac{H_s}{H_c} = 1 + \frac{\lambda_0}{2a}[1 + \tfrac{1}{2}f(\kappa)], \quad a \gg \lambda_0 \tag{V.22}$$

where $f(\kappa)$ is the same function of $\kappa$ which appears in equation V.16, and is very small for small values of $\kappa$.

For very thin films, $a \ll \lambda_0$, G-L find that

$$\left(\frac{H_s}{H_c}\right)^2 = 6\left(\frac{\lambda_0}{a}\right)^2 - \frac{7}{10}\kappa^2 + \frac{11}{1400}\left(\frac{a}{\lambda_0}\right)^2 - \ldots, \qquad (V.23)$$

which for very small $\kappa$ reduces to

$$\frac{H_s}{H_c} = \sqrt{6}\frac{\lambda_0}{a}. \qquad (V.24)$$

The same expression gives the supercooling field for thicker films (see Section 6.4). Extensive experimental verification of the temperature as well as the thickness dependence of $H_s$ has been obtained by Sevastyonov (1961, 1962) and by Khukareva (1961, 1963).

Expressions similar to V.22 and V.24 have also been derived for spheres and wires (Silin, 1951; Ginzburg, 1958a; Hauser and Helfand, 1962), and have been used by Lutes (1957) in the interpretation of his measurements of the critical field enhancement in tin whiskers.

It is possible to relate the thin film critical field to the basic superconducting parameters $\xi_0$ and $\lambda_b$ of the bulk material. The penetration depth appearing in VII.8 should be given by IV.20, in which $\xi(l)$ is determined by IV.13 with the film thickness taken as the effective mean free path (Tinkham, 1958). In the limit $a \ll \xi_0$ this yields

$$\frac{H_s}{H_c} = \sqrt{3}\left(\frac{\lambda_b^2 \xi_0}{a^3}\right)^{1/2} \qquad (V.25)$$

(Douglass and Blumberg, 1962). The use of the thin film susceptibility as derived by Schrieffer (1957) with non-local electrodynamics leads to 20–40 per cent higher values of the numerical constant (Ferrell and Glick, 1962; Toxen, 1962). This has been studied in great detail by DeGennes and Tinkham (1964). Guyon *et al.* (1967) have made a complete study of stability conditions and of the order of the transition.

The size effect on the critical field is particularly striking in experiments using extremely thin evaporated films. In their experiment on the Knight shift in tin, Androes and Knight (1961) used films of thickness $\approx 100$ Å and found $H_c(0) \approx 25$ kgauss. Ginsberg and Tinkham (1960) saw no effect on the superconducting properties of their 10–20 Å lead film in a field of 8 kgauss.

The equivalent of small superconducting specimens can exist also in bulk material. In an inhomogeneous specimen there will be local variation of the surface energy due to varying strain or to varying electronic mean free path. If locally the surface energy is sufficiently lower than the value elsewhere, it may be energetically favourable for this region to remain superconducting in the presence of an external field even when the surrounding material has become normal (Pippard, 1955). Under these conditions one can thus have a situation quite analogous to that of small specimens: small superconducting regions exist in a matrix of normal material (Gorter, 1935; Mendelssohn, 1935; Shaw and Mapother, 1960). If their dimensions are small compared to the penetration depth, the critical field of these regions will be correspondingly raised, and it is known (Faber and Pippard, 1955b; Cochran *et al.*, 1958) that such regions can persist in high fields. In many instances these regions are threads which can form continuous superconducting paths from one end of the specimen to the other, resulting in a resistive transition much broader and extending to much higher fields than the magnetic one (Doidge, 1956). The threads are, of course, likely to touch each other in many places, resulting in what Mendelssohn (1935) called a superconducting sponge. The multiple connectivity of such a structure generally leads to highly irreversible magnetic transition with almost total flux trapping. Bean (1962) has used a simplified model with which to calculate the magnetization curve of such a sponge. He has confirmed some features of this model with an artificial filamentary superconductor made by forcing mercury into the pores of Vycor glass (Bean, 1964).

### 5.3. Variation of the order parameter and the energy gap with magnetic field

From equations V.G-L1 and V.G-L2 one can also calculate the variation of the order parameter $\psi$ inside the thin films. For thicknesses $2a$ very small compared to the width of the transition layer $\lambda_0/\kappa$, or in the equivalent Pippard terms for $2a \ll \xi_0$, $\psi$ can be considered constant, and one can take $\kappa \approx 0$. This leads to (Ginzburg, 1958a)

$$\frac{\psi^2(H_e)}{\psi_0^2} = \left[1 - \left(\frac{H_e}{H_c}\right)^2 \frac{a^2}{6\lambda_0^2}\right] \Big/ \left[1 - \left(\frac{2a}{\lambda_0}\right)^2 - \frac{1}{30}\left(\frac{H_e^2}{H_c^2}\right)\left(\frac{2a}{\lambda_0}\right)^4\right]. \quad \text{(V.26)}$$

For very thin films VII.8 applies, so that

$$\frac{\psi^2(H_e)}{\psi_0^2} = \left[1 - \left(\frac{H_e}{H_s}\right)^2\right]\bigg/\left[1 - \frac{4}{5}\left(\frac{2a}{\lambda_0}\right)^2\right]. \tag{V.27}$$

For such films, therefore, $\psi(H_s) = 0$, which means that the transition into the normal state is of second order, without a latent heat and with a discontinuity only in the specific heat, and not in the entropy. There can be no supercooling, and therefore, no hysteresis. For thicker films and bulk samples the transition in an external field, as discussed in Chapter II, is always of first order. The critical thickness below which there is a second order transition is

$$2a = \sqrt{(5)}\lambda_0,$$

which has been verified by Zavaritskii (1951, 1952). Note that as the penetration depth is inversely proportional to $\psi$, $\lambda(H)$ for thin films is much larger than $\lambda_0$ even in fairly small fields (Douglass, 1961c).

Douglass (1961a) has pointed out that because of the proportionality of the energy gap to $\psi$, as derived by Gor'kov (1959, 1960), equations VII.9 and VII.10 represent the field dependence of the energy gap in sufficiently thin films. Thus one can write

$$\frac{\epsilon^2(H_s)}{\epsilon_0^2} = 0 \quad \text{for } 2a < \sqrt{(5)}\lambda_0. \tag{V.28}$$

For thicker films, V.26 and V.27 do not apply, and G-L1 and G-L2 must be solved numerically. The resulting variation of the energy gap at $H_e = H_s$ as a function of film thickness has been calculated by Douglass (1961a). It is displayed by the curve in Figure 16. The points are gap values which Douglass (1961b) obtained from tunnelling experiments (see Section 10.6). Similar results have been found by Giaever and Megerle (1961), also by means of the tunnel effect, as well as by Morris and Tinkham (1961) with thermal conductivity measurements. With $H_e \approx H_s$, the empirical variation of the energy gap with field closely agrees with the Ginzburg-Landau-Gor'kov predictions even at temperatures well below $T_c$. In such high fields the order parameter is then small enough to make tenable the basic G-L assumptions as well as Gor'kov's identification of the energy gap with $\Psi$. More extensive tunneling measurements have been reported by Meservey and Douglass (1964), and by Collier and Kamper (1966).

Bardeen (1962) has calculated the critical field and critical current for thin films on the basis of the BCS theory. At higher temperatures his results generally confirm the predictions of the Ginzburg-Landau theory, including the vanishing of the energy gap and a resulting second-order transition at the critical field in sufficiently thin films. At much lower temperatures, however, below about $T_c/3$, Bardeen finds that for any thickness the energy gap remains finite and the transition a first-order one. However, Maki (1963) as well as Nambu

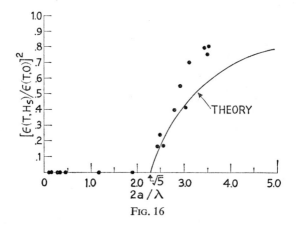

Fig. 16

and Tuan (1963) predict that the phase transition should be of the second order at all temperatures. Meservey and Douglass (1964) verify this down to $t = 0.14$.

### 5.4. Extensions of the G-L theory

The domain of validity of the G-L equations was originally thought to be limited to temperatures very near $T_c$. In this region the order parameter (and therefore the energy gap) is very small and its spatial variation very slow. However, the reformulation of the G-L equations in terms of microscopic quantities (Gor'kov 1958, 1959, 1960), has made possible a considerable and highly successful effort to extend their range to all temperatures. The usefulness of this will become apparent in Chapter VII, when applications of the G-L theory will be further discussed. This section will summarize the formal work which has been done on its extensions.

This has generally proceeded in one of two directions, each equivalent to relaxing one of the two limiting conditions mentioned above: the smallness of the energy gap, and the slowness of its variation. To allow the gap to grow large but to restrict its variation implies relatively large values of the coherence length $\xi$ (see Section 4.3). The initial work in this direction was therefore done in the limit of very pure materials (Werthamer, 1963a; Tewordt, 1963, 1965a; Zumino and Uhlenbrock, 1964; Eilenberger, 1965). Werthamer (1964) and Tewordt (1965b) extended this to finite mean free paths. The limit in which the order parameter approaches its equilibrium value has been studied by Tsuzuki (1964) in the pure and by Melik-Barkhuderov (1964) and Maki (1964b) in the impure case. The most complete method for extending the G-L theory to arbitrary values of the gap and order parameters has been developed by Eilenberger (1966).

In essence, the work just described allows the extension of the G-L theory to arbitrary temperatures in the low field limit. By contrast, the second major theoretical effort has been in the limit of high fields near the transition point, where the variations of the order parameter may be large but its value remains small. The Gor'kov formulation of the G-L relations results in an integral equation in terms of Green's functions. Maki (1964a) and De Gennes (1964) showed independently that in the dirty limit ($l \ll \xi_0$), the equations reduce to a soluble differential form. This was further elaborated by Caroli *et al.*, (1966) and by Maki (1966).

The differential formulation is necessarily a local one, so that this approximation is implicit in all the extensions discussed. Non-local electrodynamics requires the retention of integral relations which can generally be solved only for specific geometries or special limiting cases. Helfand and Werthamer (1966) and Eilenberger (1967) retain non-locality in their calculations in or very near the limit of vanishing order parameter in bulk samples. Non-local theories for thin films have been developed by Thompson and Baratoff (1965) and by

Shapoval (1965). Schattke (1966) has studied the non-local low field limit.

The original G-L theory and all the extensions mentioned thus far deal with time-independent situations in which the superconductor is in thermal equilibrium in a constant magnetic field. There has been much recent interest in attempts to describe cases in which the order parameter varies with time. This has proved to be very difficult. The most detailed work to date is that of Abrahams and Tsuneto (1967) and of Lucas and Stephen (1967).

# CHAPTER VI

# The Surface Energy

## 6.1. The surface energy and the range of coherence

Closely tied to the range of coherence of the superconducting wave functions is the existence of an appreciable surface energy on a boundary between the superconducting and normal phases. H. London (1935) already pointed out that the total exclusion of an external field does not lead to a state of lowest energy for a superconductor unless such a boundary energy exists. In the presence of an excluded external field, $H_e$, the energy of a superconductor increases by $H_e^2/8\pi$ per unit volume. It would, therefore, be energetically more favourable for a suitably shaped superconductor to divide up into a very large number of alternately normal and superconducting layers such that the width of the latter is less than $\lambda$, and that of the former very much smaller than that. The resulting penetration of the external magnetic field into the superconducting layers much reduces the magnetic energy of the sample, while the extreme narrowness of the normal layers keeps negligible their contribution to the total free energy. This situation is made energetically unfavourable by the existence of a surface energy. To make each superconducting layer narrower than $\lambda$, a slab of thickness $d$ must have $d/\lambda$ such layers. This is avoided by an interphase surface energy $\alpha_{ns}$ per unit surface whose contribution exceeds the gain in magnetic energy, that is:

$$\frac{2d}{\lambda}\alpha_{ns} > \frac{H_c^2 d}{8\pi}, \qquad \text{(VI.1)}$$

where the energies have been calculated for a volume of slab of unit surface area. Hence

$$\alpha_{ns} > \frac{\lambda}{2}\frac{H_c^2}{8\pi}. \qquad \text{(VI.1')}$$

It is convenient and customary to express the surface energy in terms of a parameter $\Delta'$ of dimensions of length, such that

$$\alpha_{ns} \equiv \Delta' \frac{H_c^2}{8\pi}. \qquad \text{(VI.2)}$$

Thus one sees that

$$\Delta' > \frac{\lambda}{2} \qquad \text{(VI.3)}$$

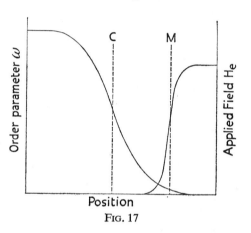

FIG. 17

is the condition for the diamagnetic behaviour of superconductors.†
Empirical values of $\Delta'$ for pure superconductors turn out to be an order of magnitude larger than the penetration depth.

The surface energy is intimately related to the Pippard range of coherence. Figure 17 shows the variation of the order parameter $\mathscr{W}$ and of the externally applied field $H_e$ along a direction perpendicular to the *s–n* interphase boundary. One can define two effective boundaries, indicated by $M$ and $C$. $M$ is the magnetic boundary defined so that if inside the superconductor $B = H_c$ up to $M$ and then dropped off sharply to zero, the total magnetic energy would equal the actual value, given by the integral of $BH/8\pi$ over the entire superconductor.

† F. London ([2], pp. 125–130) has shown that taking into account the detailed field penetration leads to the condition $\Delta' > \lambda$.

Similarly $C$ is the configurational boundary such that if $\mathscr{W}$ dropped sharply to zero at $C$ after being constant up to that point, one would have the same superconducting free energy as the actual amount. The free energy per unit volume of the superconductor is lower than that in the normal state by an amount $H_c^2/8\pi$. A configuration boundary as shown on the inside of the magnetic boundary is essentially equivalent to a reduction of the superconducting volume and hence an increase in the total free energy by an amount equal to $H_c^2/8\pi$ times the distance $C$–$M$ per unit area of interphase boundary. The introduction of the Pippard range of coherence thus leads to a configurational boundary surface energy $\Delta' \approx \xi$. From this one must subtract the decrease in energy due to the penetration of the field. Figure 17 indicates that the distance $C$–$M$ corresponds to the resulting net surface energy parameter

$$\Delta \approx \xi - \lambda. \qquad (VI.4)$$

The condition for the Meissner effect is that $\xi > \lambda$, i.e. that $\Delta > 0$.

The Ginzburg-Landau theory was formulated so as to lead explicitly to the existence of a surface energy, which arises as in the Pippard approach from the gradual variation of the order parameter $\psi$ over a finite distance, from the zero value in the normal region to its full equilibrium value in the superconducting domain. Again the surface energy is that amount which is needed to equate the energies of the two phases in equilibrium, with $H_e = H_c$. In the superconducting phase the increase in the free energy in the region where $\psi$ is changing is given by V.6; in addition there is a reduction of the energy due to the penetration of the field equal to

$$-MH_c = -\left\{\frac{H(z)-H_c}{4\pi}\right\} H_c, \qquad (VI.5)$$

where $H(z)$ is the value of the penetrated field at any point inside the superconducting region. Thus the surface energy is given by the integral of the difference between the superconducting and normal free energies over the entire superconducting half-space:

$$\alpha_{ns} = \int\limits_0^\infty dz \left\{ G_s(H,z) - \frac{H(z)H_c}{4\pi} + \frac{H_c^2}{4\pi} - G_n(0) - \frac{H_c^2}{8\pi} \right\}, \qquad (VI.6)$$

which gives for $\Delta$:

$$\Delta = 4 \int\limits_{-\infty}^{\infty} dz \left\{ \frac{\lambda_0^2}{\kappa^2} \left[ \frac{d}{dz} \left( \frac{\psi}{\psi_0} \right) \right]^2 + \left[ \frac{H(z)}{\sqrt{2}H_c} \right]^2 - \frac{H_e H(z)}{2H_c^2} \right\}, \quad \text{(VI.7)}$$

where $\lambda_0$ is the empirical penetration depth into a bulk superconductor, and $\kappa$ the dimensionless parameter defined in the previous chapter. This equation requires numerical integration. For $\kappa \ll 1$ it reduces to

$$\Delta = 1 \cdot 89 \frac{\lambda_0}{\kappa}. \quad \text{(VI.8)}$$

The thickness of the transition layer is thus, according to the G-L theory, of the order of $\lambda_0/\kappa \approx 10\lambda_0$ for most pure elements. The intimate relation between the G-L model and Pippard's range of coherence is shown by Gorkov's derivation of G-L1 and G-L2 from first principles. He finds an expression for the G-L parameter $\kappa$ in terms of the critical temperature and the Fermi momentum and velocity of the metal. Using equations IV.17 and III.13, this simplifies to

$$\kappa \approx 0 \cdot 96 \frac{\lambda_L(0)}{\xi_0}. \quad \text{(VI.9)}$$

Comparing VI.8 and VI.9 shows that, as expected, the Pippard range of coherence $\xi_0$ and the surface energy parameter $\Delta$ as derived from the G-L theory are of comparable size. In short, both approaches necessarily lead to a positive surface energy because both require that the characteristic superconducting order parameter vary over a finite distance. Both, therefore, obtain a net surface energy parameter of length comparable to the difference between this distance and the penetration depth of an external magnetic field.

It therefore also follows from both theories that the surface energy must decrease and may even become negative when the range of coherence decreases and the penetration depth increases. Equations IV.12 and IV.21 show that this is just what happens to $\lambda$ and to $\xi$ when the mean free path of the superconductor decreases. In alloys one would, therefore, expect $\Delta$ to decrease with increasing impurity content, and ultimately to become negative. This has indeed been inferred by Pippard (1955) and by Doidge (1956) from their studies of flux trapping and the superconducting transition in dilute solid solutions of indium in tin. Direct measurements of $\Delta$ in such alloys by Davies (1960) has demonstrated its decrease with shortening $l$, and Wipf (1961) has traced this decrease to actual negative values. All the work cited indicates that $\Delta$ becomes negative at a critical concentration of approximately 2·5 atomic per cent of indium in tin.

Changes of $\Delta$ with decreasing mean free path also follow from the numerical integration of VI.7, which yields that $\Delta \leqslant 0$ for

$$\kappa \geqslant 1/\sqrt{2}. \tag{VI.10}$$

This prediction is in good agreement with the work on tin–indium alloys just cited. Chambers (1956) found that the addition of 2·5 atomic per cent of indium to tin about doubles the penetration depth as compared to pure tin, so that according to the defining equation $\kappa$ should be increased by a factor of approximately four. This would make $\kappa \approx 0.6$, which is close to the theoretical value of 0·707.

### 6.2. The surface energy and the intermediate state

Chapter III mentioned that a superconducting specimen with a demagnetization coefficient $D$ is in the intermediate state when the external magnetic field $H_e$ satisfies the inequality $(1-D)H_c < H_e < H_c$. All experiments on the detailed structure of this state have generally substantiated the suggestion of Landau (1937, 1943) of a laminar structure of alternating normal and superconducting layers. The thickness of the normal layer grows at the expense of the superconducting one as the external field approaches $H_c$. Landau further suggested that in the normal layers $B = H_c$, while $B = 0$ in the superconducting ones.

Clearly the width of the laminae is strongly influenced by the magnitude of the interphase surface energy $\Delta$. Indeed Landau finds that for an infinite plate of thickness $L$ oriented perpendicularly to the field $(D = 1)$, the sum $a$ of the thickness of the superconducting layer, $a_s$, and that of the normal one, $a_n$, is given by

$$a = \left( \sqrt{\frac{L\Delta}{\Psi}} \right), \tag{VI.11}$$

where $\Psi$ is a complicated function of the ratio of the external to the critical field $H_e/H_c$. Numerical values for $\Psi(H_e/H_c)$ have been calculated by Lifshitz and Sharvin (1951). A typical result is a value $a \approx 1.4$ mm for $L = 1$ cm and $H_e/H_c = 0.8$. A similar equation has also been derived by Kuper (1951), who predicts numerical values which are smaller by a factor of two or three. Typical experimental results fall in between these predictions.

These results have been obtained by a variety of methods, all making use of the fact that in the intermediate state lines of flux pass only through the normal laminae, and emerge from the specimen wherever these laminae end on the surface. A number of authors (Meshkovskii and Shalnikov, 1947; Shiffman, 1960, 1961) have

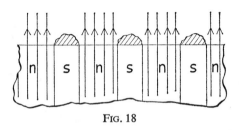

FIG. 18

passed very fine bismuth wire probes across the surface of a specimen, and observed the magnetoresistive fluctuations in the probe resistance when passing from the end of a normal lamina to that of a superconducting one. Others have spread on the surface of a flat specimen fine powder, superconducting (Schawlow *et al.*, 1954; Schawlow, 1956; Faber, 1958; Haenssler and Rinderer, 1960) or ferromagnetic (Balashova and Sharvin, 1956; Sharvin, 1960). The former will shun flux and cluster on the ends of the superconducting laminae, as shown schematically in Figure 18; the latter will be attracted by flux and move onto the ends of the normal laminae. The resulting powder patterns can be easily seen and photographed. Another optical method consists of placing a thin sheet of magneto-optic glass (for example, cerium phosphate glass) on the specimen surface, and observing the reflection of polarized light (P. B. Alers, 1957, 1959; DeSorbo 1960, 1965). Goodman and Werthamer (1965) and Goodman *et al.* (1966) have used high-speed photography to observe extremely rapid motion of flux patterns.

The frontispiece shows a series of photographs obtained by Faber (1958) with superconducting tin powder on an aluminium plate, taken with increasing external field oriented perpendicularly. The dark areas are covered with powder and are therefore the ends of the superconducting laminae. The gradual shrinking of these areas with increasing field and the corresponding growth of the light, normal regions is clearly visible. The domains show a peculiar type of corrugation, not predicted by the Landau model, and adding to the surface to volume ratio of the laminae.

Work with these various techniques has been continued in recent years, with interest focussing particularly on dynamic effects associated with the current-carrying intermediate state (see, e.g., Chandrasekhar *et al.*, 1966; Sharvin, 1965, 1966; Haenssler and Rinderer, 1966).

### 6.3. Phase nucleation and propagation

H. London (1935) pointed out that the existence of a positive surface energy at the interphase boundary must under suitable conditions give rise to phenomena analogous to superheating and supercooling in the more familiar phase transitions. In fact a stable nucleus for the phase transition cannot exist at all if the surface energy is everywhere positive. Indeed there are many experimental observations that when a specimen is placed in a greater than critical magnetic field which is then reduced, the normal phase persists in fields less than $H_c$. This is the superconducting equivalent of supercooling. A typical magnetization curve illustrating this is shown in Figure 19. The degree of this 'supercooling' is characterized by the parameter $S_l \equiv H_l/H_c$, or by the parameter

$$\phi_l \equiv 1 - S_l^2 = (H_c^2 - H_l^2)/H_c^2. \tag{VI.12}$$

For tin, $S_l$ is commonly of the order of 0·9; in aluminium the degree of supercooling is usually much larger, and values of $S_l$ as low as 0·02 have been observed.

Superheating is the name given to the persistence of the superconducting phase at fields above $H_c$. This is very rarely observed.

Garfunkel and Serin (1952) have shown that this is so because the ends of any conventional specimen cannot resist the initiation of the normal phase, probably because of large local field values resulting from demagnetization effects. Centre portions of long tin rods could be made to superheat to $S_l = 1.17$.

Much information on the nucleation of the superconducting phase and on its relation to the surface energy has been obtained by Faber

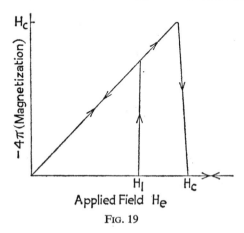

FIG. 19

(1952, 1955, 1957) in a series of measurements on supercooling in tin end aluminium. His technique consisted of winding on a long cylindrical specimen several small, spaced coils the field of which could be made to add or to subtract from a field produced by a large solenoid surrounding the entire sample. With the sample normal, the field of the solenoid could be lowered to some value between $H_l$ and $H_c$, and the field could then be lowered locally by a suitably directed current through one of the smaller coils. The superconducting phase then nucleated in the portion of the sample under the coil, and spread rapidly throughout the sample. In this fashion supercooling could be studied at different portions of the sample. The transition was detected by pick-up coils distributed along the specimen.

At a given temperature the degree of supercooling varied considerably from point to point in a given specimen but at a given point frequently remained reproducible even when in between measurements the specimen was warmed to room temperature. This indicated that nucleation must occur at particular spots, some of which promote nucleation more effectively than others. As the surface energy can be lowered and may even become negative due to strain, it is reasonable to assume that the spots favouring nucleation are regions of local strain, some of which exist in even the purest specimens. This is supported by Faber's finding that any handling of the specimens between measurements could change the location and effectiveness of the nucleation centres. Strained regions probably contain a high density of dislocations.

By correlating the size of the nucleating field $H_l$ with the time it took to be effective, Faber could deduce the depth of the nucleating flaw below the sample surface, and found this always to be between $10^{-4}$ and $10^{-3}$ cm. Etching down the surface to this depth would uncover further flaws extending to a similar depth. It is thus reasonable to take $10^{-4}$–$10^{-3}$ cm as being the approximate size of the nucleating flaws. At temperatures well below $T_c$, this length is considerably bigger than the width of the interphase boundary, and one can therefore imagine such a flaw to consist of a region of negative surface energy surrounded by a shell across which the surface energy increases to the normal positive value of the bulk material. Faber (1952) has shown that there is a potential barrier against the further growth of this nucleus until one has reached a degree of supercooling such that

$$\phi_l \approx \frac{\Delta}{r} + n, \tag{VI.13}$$

where $\Delta$ is the surface energy parameter, $r$ a length of the order of the flaw size, and $n$ a small constant determined by the flaw's shape and demagnetization factor. The measurements in fact show that the temperature variation of $\phi_l$ is very much like that of $\Delta$, as determined from other experiments.

Both Faber (see Faber and Pippard, 1955b) and Cochran *et al.* (1958) found that supercooling was much enhanced after a specimen had been placed temporarily in a field much higher than the bulk critical value. This shows that certain superconducting nuclei can be

6

quenched only by such a high field and supports much other evidence that in a non-ideal specimen there can exist small regions of high strain which remain superconducting in very high fields.

By means of a series of pick-up coils along his specimens, Faber (1954) was able to observe the propagation of the superconducting phase once the transition had been initiated at some nucleus. From his results he infers that the growth of the superconducting phase occurs in a series of distinct stages. The nucleus, which is always near the surface, first expands to form an annular sheath around the specimen. This sheath then spreads along the entire length of the specimen with a velocity of the order of 10 cm/sec, and finally the superconducting phase spreads inwards to fill the entire sample.

The growth of a superconducting region is limited principally by the interphase surface energy on the one hand, and by eddy current damping on the other. If there were no surface energy, the super-conducting phase could propagate by means of very thin filaments which displace no magnetic flux and therefore create no retarding induced currents. For a sheath of finite thickness, on the other hand, which propagates in the presence of an external field $H_e$, eddy currents are generated, and the magnetic energy gained in the phase transition is balanced by the unfavourable surface energy as well as the eddy current joule heating. Faber (1954, 1955) has shown that the resulting velocity of propagation for very pure specimens under optimum conditions is given by

$$v = A(l/\sigma)\Delta^{-2}([H_c - H_e]/H_c)^3, \qquad (VI.14)$$

where $l$ is the electronic mean free path in the normal phase, $\sigma$ the normal electrical conductivity, and $A$ is a constant of the specimen. By measuring the temperature variation of $v$, Faber has used this equation to obtain the temperature variation of $\Delta$ for tin and for aluminium.

The values of $\Delta(T)$ obtained in this way by Faber, as well as those measured in different ways by Davies (1960), Sharvin (1960), and Shiffman (1960), can be fitted by a number of empirical functions of temperature. According to the G-L theory, the surface energy should have the same temperature variation as $\lambda_0$, at least very near $T_c$, where $\kappa$ is independent of temperature. Hence one would expect

$$\Delta(t) = \Delta(0)(1-t^4)^{-1/2}, \qquad (VI.15)$$

which can also be written

$$\Delta(t) \approx \frac{\Delta(0)}{2}(1-t)^{-1/2} \quad \text{for } t \approx 1. \qquad (VI.15')$$

The second of these functions appears to give a good fit to various results for tin over a rather wide range of temperature, but Faber's aluminium data can be represented only by the first of these. There seems to be a definite difference in the temperature dependence of the surface energy for these two metals which is at present not understood.

The uncertainty in the temperature dependence of $\Delta$ of course introduces a degree of doubt about the extrapolated value at 0°K. The table below lists the best available values of $\Delta(0)$ for a number of metals, from a comparison of all available experimental data. Also listed are values of $\xi_0$, the range of coherence, as calculated from equation IV.17, as well as empirical values for $\lambda_0(0)$.

| *Element* | $10^5 \, \Delta(0)$ (cm) | $10^5 \, \xi_0$ (cm) | $10^5 \, \lambda_0(0)$ (cm) |
|---|---|---|---|
| Aluminium | 18 | 16 | 0·50 |
| Indium | 3·4 | 4·4 | 0·64 |
| Tin | 2·3 | 2·3 | 0·51 |

## 6.4. Supercooling in ideal specimens

Near $T_c$, $\Delta$ becomes large, and the flaws lose their effectiveness as nucleation centres. Measuring $H_l$ in this region can, therefore, give some information on supercooling in ideal, unflawed material. Faber (1957) has found for aluminium, $S_l = 0·036$, for In 0·16, and for Sn 0·23; values of Cochran *et al.* (1958) for aluminium are in reasonable agreement. These results can be compared with theoretical predictions arising from the G-L model. Equation V.9' has an interesting consequence with regard to the normal phase. One would expect that with $H_e \geqslant H_c$, the half-space described by the equation would be entirely normal, with $\psi = 0$. This is indeed a solution, but the equation is also satisfied by a second solution with $\psi \neq 0$. Assuming that for this solution $\psi/\psi_0 \ll 1$, so that $H(z) \approx H_e$ everywhere, and remembering that in the geometry chosen $A(z) = H(z)z$, the equation becomes

$$\frac{d^2\psi}{dz^2} \approx -\frac{\kappa^2}{\lambda_0^2}\left(1 - \frac{H_e^2 z^2}{2H_c^2}\right)\psi. \tag{VI.16}$$

This has the form of a wave equation for a harmonic oscillator, which is known to have periodic solutions $\psi$ which vanish for $z = \pm\infty$ (which is the required boundary condition for the normal phase) if

$$\kappa = \sqrt{2}\frac{H_e}{H_c}(n+\tfrac{1}{2}), n = 0, 1, 2, \ldots$$

In other words, for any value of $\kappa$, the normal phase of the superconductor becomes unstable with regard to the formation of laminae of superconducting material when

$$H_e/H_c = \kappa/(n+\tfrac{1}{2})\sqrt{2}, \tag{VI.17}$$

of which the highest value, with $n = 0$, is

$$H_{c2}/H_c = \sqrt{(2)}\,\kappa. \tag{VI.18}$$

A distinction must now be made as to whether $\kappa < 1/\sqrt{2}$ or $\kappa > 1/\sqrt{2}$. In the former case, which is that of most pure superconductors, $H_{c2} < H_c$, and the field $H_{c2}$ is then the lowest field to which the normal phase can persist in a metastable fashion. $H_{c2}$ is thus the lower limit to which an ideal superconductor can be supercooled, and therefore in the region very near $T_c$ one would expect the experimental value of $S_l$ to equal $\sqrt{(2)}\,\kappa$ (Ginzburg, 1956, 1958a; Gor'kov, 1959b, c).

The values of $\kappa$ calculated in this fashion from Faber's measurements of $S_l$ are: 0·164 for tin, 0·112 for indium, and 0·026 for aluminium. The first two of these agree very well with $\kappa$ values deduced from experimental penetration depths. In aluminium the lack of agreement is probably due to the appearance of non-local effects very close to $T_c$. Ginzburg (1958b) has noted that this is more likely to invalidate calculations involving the penetration depth than those regarding the surface energy and supercooling. Non-local effects become important for the former when $\xi_0 \geqslant \lambda_0$; for the latter only when

$$\xi_0 \geqslant \frac{\lambda_0}{\kappa} \gg \lambda_0.$$

Thus $\kappa$-values calculated from supercooling data are probably the most reliable, except for the effect to be discussed in Section 7.7.

For superconductors with a dimension small compared to $\lambda_0/\kappa \approx \xi_0$, the order parameter is essentially constant throughout and one can solve the G-L equation with the simplifying assumption $\kappa \approx 0$. The critical fields of supercooling are then given by

$$H_{c2} = \sqrt{6}\frac{\lambda_0}{a}H_c$$

for a slab of thickness $2a$,

$$H_{c2} = 2\sqrt{5}\frac{\lambda_0}{a}H_c$$

for a sphere of radius $a$, and

$$H_{c2} = \sqrt{8}\frac{\lambda_0}{r}H_c$$

for a cylinder of radius $r$ (Ginzburg, 1958a). For all these geometries $H_{c2}$ decreases with increasing specimen size, approaching monotonically the value given by VI.18, which depends only on the value of $\kappa$ characteristic of the material.

The compatibility of the G-L theory with the Pippard range of coherence under those circumstances of temperature, size, or mean free path which eliminate the need for a non-local electromagnetic formulation is brought out once again by the similarity of VI.18 with the corresponding expression derived by Pippard (1955). He finds, also by minimizing the free energy, that

$$H_{c2} = \frac{2\sqrt{3}}{\pi}\frac{\lambda_0}{\xi}H_c. \tag{VI.19}$$

This differs from VI.18 only by a numerical factor of order unity since $\kappa \approx \lambda_0/\xi$.

# CHAPTER VII

# Type II Superconductors

## 7.1. Basic properties near $T_c$

According to equation VI.7, the surface energy becomes negative when $\kappa > 1/\sqrt{2}$. A similar conclusion follows from the Pippard non-local model when $\lambda > \xi$ (equation VI.4: see Doidge, 1956). The existence of a positive surface energy was shown to be necessary for much of the magnetic behaviour usually found in superconductors. It is, therefore, not surprising that superconductors in which this energy is negative display quite different characteristics. They are accordingly called *superconductors of the second kind* or *type II superconductors*.

For a bulk specimen of such a superconductor the *volume* free energy in the superconducting phase remains lower than that of the normal one in external fields up to the thermodynamic value $H_c$ defined by equation II.4. The negative *surface* energy, however, makes it energetically favourable for interphase boundaries to appear at field lower than $H_c$, and for superconducting regions to persist to fields higher than $H_c$. Goodman (1961) has shown that this can already be deduced from the London model by the single addition of a negative surface energy term.

The details of the behaviour of superconductors of the second kind can be deduced from the G-L theory, which is equally valid for $\kappa > 1/\sqrt{2}$ as for $\kappa < 1/\sqrt{2}$. In particular, the analysis of the preceding chapter still holds; that is, the normal phase has a stability limit at a field $H_{c2}$ given by equation VI.18, which shows that for $\kappa > 1/\sqrt{2}$, $H_{c2} > H_c$. Abrikosov (1957) has used the G-L equations to analyze in some detail the magnetic behaviour of superconductors of the second kind, and finds the features indicated in the magnetization curve shown in Figure 20 for a cylindrical sample parallel to an external field $H_c$. There is a complete Meissner effect only up to $H_e = H_{c1} < H_c$, at which point the magnetization changes with infinite slope. For values of $\kappa$ not much larger than $1/\sqrt{2}$, Abrikosov predicts in fact a discontinuity. At somewhat higher field, the magnetization approaches zero linearly, with a slope

$$-4\pi\frac{dM}{dH} = 1/1{\cdot}18(2\kappa^2-1),\qquad\text{(VII.1)}$$

and vanishes entirely at

$$H_e = H_{c2} = \sqrt{(2)}\kappa H_c.\qquad\text{(VI.18)}$$

The magnetization curve should be fully reversible. Abrikosov cannot solve the equations determining $H_{c1}$ for all values of $\kappa$; for $\kappa \gg 1$ he obtains

$$\sqrt{(2)}\kappa H_{c1}/H_c = \ln\kappa+0{\cdot}08\qquad\text{(VII.2)}$$

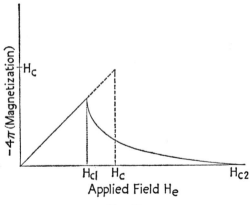

Fig. 20

In the limit $\kappa = 1/\sqrt{2}$, $H_{c1} = H_c = H_{c2}$. Goodman (1962a) has interpolated between the latter value and those given by VII.2 to get a graphical representation of $H_{c1}/H_c$ for all $\kappa$. A numerical solution has been obtained by Harden and Arp (1963).

The magnetization curve predicted by the Ginzburg-Landau-Abrikosov (G-L-A) model can be compared with experiment, as it is possible to determine the value of $\kappa$ for a specimen by independent measurement. Gor'kov (1959) has derived an expression for $\kappa$ valid when the electronic mean free path is much smaller than the intrinsic coherence length $\xi_0$. This was shown by Goodman (1962a) to have the convenient form

$$\kappa = \kappa_0 + 7 \cdot 5 \times 10^3 \gamma^{1/2} \rho. \qquad \text{(VII.3)}$$

$\kappa_0$ is the parameter for the pure substance, $\gamma$ the Sommerfeld specific heat constant, in erg $cm^{-3}$ $deg^{-2}$, and $\rho$ the residual resistivity in ohm-cm. For tin this predicts quite closely the resistivity at which the surface energy becomes negative (Chambers, 1956).

Using this equation, Goodman (1962a) has shown that the G-L-A model satisfactorily explains the magnetic behaviour of substances such as Ta–Nb alloys investigated by Calverley and Rose-Innes (1960) and his own U–Mo alloys (Goodman *et al.*, 1960). Furthermore, recent magnetization measurements on Pb–Tl single crystals (Bon Mardion *et al.*, 1962), indicate a considerable degree of reversibility. Detailed quantitative verification of the G-L-A magnetization curves, as is possible only near $T_c$, was provided by Kinsel *et al.* (1962), who used In–Bi specimens to compare values of $\kappa$ calculated from equations VI.18, VII.1, VII.3, and from Harden and Arp's values of $H_{c1}/H_c$. The different values of $\kappa$ for a given specimen agree to within a few per cent. Similar agreement can also be deduced from the results of Stout and Guttman (1952) on In–Tl alloys. Goodman (1964b) showed that the detailed shape of the magnetization curves of Mo–Re alloys (Joiner and Blaugher, 1964) agrees well with the theory. The G-L-A model for type II superconductors is thus well established near $T_c$.

The negative surface energy need not be due to a short mean free path. In principle, it is possible for the coherence length to be shorter than the penetration depth, even in a pure superconductor; this is most likely in superconductors with a high $T_c$ (see equation IV.17). Such substances are called *intrinsic* type II superconductors. The only known examples among the elements are niobium (Stromberg and Swenson, 1962; Autler *et al.*, 1962; Goedemoed *et al.*, 1963; Finnemore *et al.*, 1966) and vanadium (Radebaugh and Keesom, 1966a, b). Pure tantalum, on the other hand, is a type I superconductor (Buchanan *et al.*, 1965).

## 7.2. The mixed state

The magnetization curve of type II superconductors clearly shows that for $H_{c1} < H_e < H_{c2}$, the material is neither in the usual superconducting nor in the normal phase. Abrikosov (1957) has called this region the *mixed state*, and De Gennes ([14]) has suggested naming it the *Shubnikov phase*, honouring the scientist who first suggested the fundamental nature of type II superconductivity (Shubnikov *et al.*, 1937).

It is evident from the importance of the negative surface energy that in the mixed state the specimen must contain as large an area of interphase surfaces as is compatible with a minimum of normal volume. This could be brought about by a division of the material into a large number of very thin normal and superconducting sheets or laminae (Goodman, 1961, 1964; Gorter, 1964). According to the G-L-A theory, however, the mixed state consists of a regular array of normal filaments of negligible thickness which are arranged parallel to the external field and are surrounded by superconducting material. At the normal filaments the superconducting order parameter vanishes, and then rises from these linearly with distance. It reaches its maximum value as quickly as possible, that is over a distance of the order of $\xi$. The magnetic field has a maximum value at the normal filaments, and falls off over a distance of the order of $\lambda > \xi$. This means that the field decreases to zero only if the filaments are spaced at distances at least of the order of $\lambda$. This mixed state structure can be shown to have a lower energy than any laminar arrangement ([14], p. 71).

One can thus think of the mixed state as if the superconducting material were pierced by a number of infinitesimally thin filamentary holes, regularly spaced parallel to the external field and thus each containing magnetic flux. From the discussion in Chapter II it therefore follows that the total flux associated with each normal thread is quantized in units of $\phi_0$. This flux does not penetrate far into the superconducting material because of superconducting currents circulating in planes perpendicular to the filament. This creates a *vortex line* of superconducting pairs along each normal thread, in striking analogy to the vortices existing in liquid Helium II (Rayfield and Reif, 1964). The vortex lines in a superconductor are also often called *fluxoids*.

The flux and the currents associated with an isolated vortex line extend over a distance of about $\lambda$. The interaction between two vortex lines can thus be appreciable only at distances less than $\lambda$. This means that when the formation of vortex lines becomes energetically favourable at $H = H_{c1}$, they can essentially immediately achieve a density corresponding to a separation of about $\lambda$ without creating much interaction energy ([14], pp. 66ff.). This causes the abrupt decrease of the magnetization at $H_{c1}$ predicted by Abrikosov and verified experimentally. It is not certain, however, whether the magnetization actually decreases discontinuously at this field or whether it merely drops with an infinite slope. The former would correspond to a first order transition with a latent heat, the latter to a second-order transition of the $\lambda$-type, with an infinity in the specific heat.

In principle this question could be settled either by direct specific heat measurements or from determination of the magnetization, but it is very difficult to do so unequivocally. However, very careful and detailed measurements on high-purity niobium of the specific heat (McConville and Serin, 1965) and of the magnetization near $H_{c1}$ (Serin, 1965) indicate rather conclusively that ideal specimens exhibit a second-order $\lambda$-transition at $H_{c1}$.

With the external field increasing beyond $H_{c1}$, more and more vortex lines are formed until their spacing approaches

$$2\sqrt{\left(\frac{\pi}{2}\right)\frac{\lambda_0}{\kappa}} \approx \xi \qquad (\text{VII.4})$$

as $H$ nears $H_{c2}$ (Abrikosov, 1957). According to Abrikosov, the vortex lines form a square array at all fields except very near $H_{c1}$, but it has been shown (Kleiner *et al.*, 1964; Matricon, 1964) that a triangular array has a somewhat lower energy throughout the mixed state. Furthermore, unlike a square array, a triangular one is microscopically stable against small vibrational deviations (DeGennes and Matricon, 1964; Matricon, 1964; Fetter *et al.*, 1966; Fetter, 1966). For a triangular array the numerical coefficient in equation VII.1 changes from 1.18 to 1.16.

DeGennes and Matricon (1964) suggested the possibility of investigating the nature of the vortex line structure by slow neutron diffraction, and this has been done successfully by Cribier *et al.* (1964) in spite of the considerable experimental difficulties (see [14], p. 74). The results indicate unambiguously the formation of a triangular array.

This has been confirmed more recently by Essmann and Traüble (1966, 1967) who observed the pattern formed by very small cobalt particles deposited on the surface of the superconductor by evaporation. The particles preferentially lodge on the magnetized region of the surface.

One can also infer the nature of the vortex line array from the effect of the inhomogeneous field distribution on the shape of the absorption line in nuclear magnetic resonance (Fite and Redfield, 1966).

A fundamental feature of the vortex structure of the mixed state is that the order parameter $\Psi$ is everywhere finite except along the centre of the vortices, which are normal filaments of negligible volume. Thus the material can still be considered as entirely superconducting. Abrikosov (1957) showed in fact that in the mixed state one can characterize the material by a mean square order parameter $\overline{\Psi^2}$, and that near $H_{c2}$ this varied linearly with the magnetization. The correctness of this and therefore the validity of the vortex structure has been substantiated by measurements of the specific heat (Morin, *et al.*, 1962; Goodman, 1962b; Hake, 1964; Hake and Brammer, 1964) and of the thermal conductivity (Dubeck *et al.*, 1962, 1964; Lindenfeld *et al.*, 1966).

## 7.3. Properties of the vortex lines
Figure 21 illustrates what is meant by a vortex line in the mixed state. It displays the variation with radial distance $r$ from the vortex centre of the order parameter $\Psi(r)$, and the flux density, $B(r)$, and also shows qualitatively the current flow and the lines of magnetic flux in the vortex. The region over which $\Psi(r)$ varies from zero to its full value is called the *core* of the vortex.

In studying the properties of isolated vortex lines as well as their interaction (see [14], Chapter 3) it is customary to approximate the vortex line by a normal cylinder of radius $\xi$ within which the order parameter vanishes, and beyond which it has its full value. Clearly this is a good approximation if $\kappa \simeq \lambda/\xi \gg 1$; the larger $\kappa$, the more one can entirely neglect the core and treat the order parameter as essentially constant throughout the specimen.

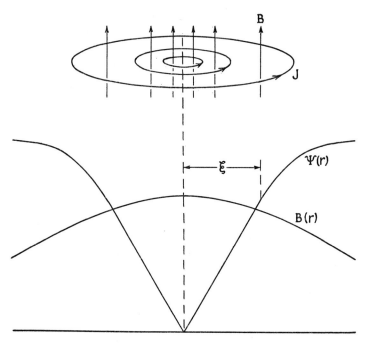

FIG. 21

Caroli *et al.* (1964) have shown that the core of the vortex line contains low-lying energy levels for single electrons, rather like in a normal metal. This further justifies treating the core as a normal region. The existence of these quasi-normal excitations in the core explains a number of experimental results in low and intermediate fields: the finite absorption of microwaves in the mixed state (Rosenblum and Cardona, 1964b), the presence of a linear term in the superconducting specific heat (Hake and Bremmer, 1964; Ferreira *et al.*, 1964, 1966; Keesom and Radebaugh, 1964b), and the low value as well as the field dependence of the nuclear spin-lattice relaxation times (Silbernagel *et al.*, 1966; Fite and Redfield, 1966). The presence of a linear specific heat term had been predicted earlier on thermo-dynamic grounds by Gorter *et al.* (1964).

Near the upper critical field $H_{c2}$ increasing overlap between neigh-bouring vortex lines drastically modifies the core states. As discussed in Section 12.4, it is then more appropriate to think of the entire material as being in a gapless state.

In a finite specimen, a vortex line near the surface interacts with the external field and the associated screening currents (see [14], pp. 76–80). This leads to a surface barrier effect inhibiting the motion of flux lines in and out of an ideal specimen, as predicted independently by Bean and Livingston (1963) and DeGennes and Matricon (1964). Resulting delayed flux entry has been observed by Joseph and Tomasch (1964), DeBlois and DeSorbo (1964), and Boato *et al.* (1965).

In an ideal type II superconductor, homogeneous and devoid of lattice imperfection, the vortex lines would be pushed out of the material by the Lorentz force if the specimen carried any current at right angles to the field (Gorter, 1962a, b). In any actual material, the motion of the lines is inhibited by defects and inhomogeneities which form potential barriers by which lines the are pinned. Anderson (1963) has investigated the thermally activated 'creep' of lines at low current densities, and has shown that on a local scale the density of lines tends to remain uniform, so that bundles of lines move together. This vortex or flux creep has been further discussed by Friedel *et al.* (1963) and by Anderson and Kim (1964). The measurements of Kim *et al.* (1963) indicated the activation energies needed to free the flux bundles from their pinning centres, and investigations by Van Ooijen and Van Gurp (1965) of the noise due to creep suggest that each bundle contains $10^3$–$10^5$ vortex lines.

With increasing current densities creep changes into a viscous flow of vortex lines, giving rise to resistive phenomena which have been extensively investigated (Anderson and Kim, 1964; Strnad *et al.*, 1964; Kim *et al.*, 1965; Staas *et al.*, 1964; Niessen *et al.*, 1965;

Swartz and Hart, 1965). Strnad *et al.* (1964) were the first to draw attention to a remarkably simple empirical relationship which appears to be followed by the mixed state resistance $\rho_f$ not too close to $T_c$:

$$\rho_f/\rho_n = H/H_{c2}. \tag{VII.5}$$

Here $H$ is the mean field inside the specimen, and $\rho_n$ the normal state resistance. This simple relation leads to an interesting physical interpretation, which can be shown as follows: If $d$ is the average distance between flux lines, each carrying one flux quantum $\phi_0$, then $H = \phi_0/d^2$. Also it follows from equation VII.4 that $H_{c2} = \phi_0/\xi^2$. Hence

$$H/H_{c2} = \xi^2/d^2. \tag{VII.6}$$

Using the simple approximation that vortices contain normal cores of radius $\xi$, the resistance ratio is thus seen to be equal to the fraction of normal material in the mixed state. The empirical relation thus suggests that the mixed state resistance is due to the resistance of the normal cores, implying that the current density is uniform throughout the mixed state. A similar inference had been drawn by Rosenblum and Cardona (1964b) to explain their surface resistance data.

Experimental studies of vortex motion by measurements of the d.c. resistance have been supplemented by a number of other methods. Much of this has concentrated on the Hall angle, the tangent of which equals the ratio between the electric fields parallel to and at right angles to the direction of current flow (Reed *et al.*, 1965; Niessen and Staas, 1965; Staas *et al.*, 1965). The complement of this angle is the angle at which the flux lines move with respect to the measuring current as a result of the Lorentz force. The present experimental evidence with regard to the behaviour of this angle in the mixed state is confusing. The most recent and very careful measurements by Druyvesteyn *et al.* (1966) and by Fiory and Serin (1967b) show a Hall angle which continues to vary with magnetic field as in the normal state, while the data of Maxfield (1967) suggest that the angle remains constant independently of field. (See footnote, p. 91.)

Borcherds *et al.* (1964) and Vinen and Warren (1967a) have inferred the mixed state resistance from the rate at which flux enters or leaves a long cylindrical specimen when the external field is changed by small but discrete amounts. Thermomagnetic effects due to the flux motion, in particular the Peltier and the Ettingshausen effects, have been studied by Fiory and Serin (1966, 1967a) and by Otter and Solomon (1966).

The theoretical understanding of vortex motion is as yet far from satisfactory. An early semi-quantitative treatment by Volger *et al.* (1964) which attempted to derive the observed d.c. resistance on the basis of normal electron drag was extended by Tinkham (1964c), who introduced a relaxation time for the return of the order parameter to equilibrium. In the limit of low temperatures his results reduce to the simple relation VII.5, and are in reasonable agreement with the data of Strnad *et al.* (1964) and of Swartz and Hart (1965). These semiquantitative treatments have been elaborated by Bardeen and Stephen (1965), Van Vijfeijken and Niessen (1965), Nozières and Vinen (1966), and Vinen and Warren (1967b). All these authors use a two-fluid model and simple hydrodynamic assumptions for the vortex flow, but differ from each other in that the first two assume the existence of local equilibrium throughout the vortex, while the latter two instead let the contact potential vanish at the vortex edges. As a result the two sets of theories lead to different predictions about the Hall angle behaviour, although they both derive the correct variation of the resistance.

These phenomenological treatments have been sharply criticized by Caroli and Maki (1967) on the basis of a fundamental inadequacy both of the two-fluid model and of the hydrodynamic approach. Instead, Caroli and Maki extend and generalize earlier attempts by Schmid (1966) and Kulik (1966) to treat the problem of vortex motion on a microscopic basis, using time-dependent G-L equations. Caroli and Maki calculate vortex motion at arbitrary temperatures near $H_{c2}$, for both clean and dirty materials, and find as a basic result that the order parameter moves with uniform velocity in the presence of an electric field. They also predict an a.c. effect which is analogous to the a.c. Josephson effect.

## 7.4. Type II superconduction at $T < T_c$

The Abrikosov theory of type II superconductors is based on the G-L equations which in their original form are rigorous only in a small temperature range near the transition temperature. However, as discussed in Section 5.4, the Gor'kov derivation of the G-L equations in microscopic formulation has stimulated a host of calculations to extend the range of their validity to low temperatures. In turn this allows extending the G-L-A theory of type II superconductors. Because of the important role of Gor'kov's work one now generally speaks of the GLAG theory.

The most important and historically the first contribution was the dirty-limit, small gap parameter theory of Maki (1964a, b) and De Gennes (1964), which is applicable at all temperatures near $H_{c2}$. This work indicated that the GLAG theory remains valid near $H_{c2}$ at all temperatures with the introduction of temperature-dependent G-L parameters $\kappa_1(T)$ and $\kappa_2(T)$, defined by the following equations:

$$\kappa_1(T) \equiv H_{c2}(T)/\sqrt{2}H_c(T) \qquad \text{(VII.7)}$$

and

$$-4\pi(dM/dH)_{Hc2} = 1/1\cdot16[2\kappa_2^2(T)-1]. \qquad \text{(VII.8)}$$

According to the calculations, $\kappa_1(T_c) = \kappa_2(T_c) = \kappa$, and $\kappa_1(0) = \kappa_2(0) = 1\cdot19\kappa$. For intermediate temperatures, both parameters increase slowly and monotonically with decreasing temperatures, $\kappa_2(T)$ always being one or two per cent smaller than $\kappa_1(T)$ (Caroli *et al.*, 1966). This is in excellent agreement with many experiments on alloys (see, e.g., Bon Mardion *et al.*, 1965), except that the predicted difference between $\kappa_1(T)$ and $\kappa_2(T)$ for $0 < T < T_c$ is too small to be observed. Guyon *et al.* (1967) found the same temperature variation of $\kappa_2(T)$ for thick dirty films. For films so thin as to contain no vortex lines, however, their measurements showed a decrease in $\kappa_2(T)$ with decreasing temperature, in accord with the original calculations of Maki (1964a) and later work by Thompson and Baratoff (1967).

The increase of $\kappa_1$ with decreasing temperature corresponds to a decrease in the surface energy, which continues to become negative when $\kappa_1 = 1/\sqrt{2}$. This was verified by Kinsel *et al.* (1964) with an In–Bi alloy specimen, which had $\kappa = 0\cdot62$ and was type I for $T/T_c > 0\cdot7$ and type II at lower temperatures. Thus any specimen with $0\cdot57 < \kappa < 1/\sqrt{2}$ will become type II at some $T < T_c$.

Helfand and Werthamer (1966) extended the Maki-DeGennes theory to arbitrary mean free paths in order to calculate the temperature and purity dependence of $H_{c2}(T)$ and therefore of $\kappa_1(T)/\kappa$. In the dirty limit they reproduce the Maki-DeGennes results, and find at all temperatures a small mean free path effect such that $\kappa_1(T)$ increases by a few per cent with increasing purity. At $T = 0$, $(\kappa_1/\kappa)_{\text{pure}}$ $= 1 \cdot 26$ as compared to $(\kappa_1/\kappa)_{\text{dirty}} = 1 \cdot 19$. The result for the pure limit agrees with that obtained by Gor'kov (1959a) using a trial solution for the GLAG integral equation at $T = 0°$K. Experimentally, however, $\kappa_1(T)/\kappa$ is found to increase much more rapidly than predicted as the temperature is decreased, rising to values as high as $1 \cdot 7$ at $0°$K (see, e.g., McConville and Serin, 1965). This discrepancy may well be due to Fermi surface anisotropy, the effects of which were necessarily ignored by the local electrodynamics of the theory. Calculation of anisotropy effects by Hohenberg and Werthamer (1967) show qualitatively a shift in the right direction.

Maki and Tsuzuki (1965) used a variational treatment, later justified and extended by an exact calculation of Eilenberger (1967), to obtain the temperature dependence of $\kappa_2(T)$ in the pure limit. In this limit the theory predicts an infinite value of $\kappa_2$ at $T = 0°$K, so that comparison with experiment cannot be made at low temperatures. At higher temperatures, however, there is qualitative agreement (see, e.g., Finnemore *et al.*, 1966) with the prediction that $\kappa_2 > \kappa_1$ and that at any temperature the difference between the two parameters decreases with decreasing purity.

Extensions of the G-L theory in the small field limit have been used by Neumann and Tewordt (1966a), Maki (1964b), and Melik-Barkhuderov (1964) to treat type II superconductors at low temperatures near $H_{c1}$. These calculations can only be carried out in the limit $\kappa \gg 1$, for which the vortex cores constitute a negligible fraction of the whole material and the order parameter is thus close to unity throughout the specimen. Under these conditions the GLAG model continues to be valid at all temperatures if one introduces a third temperature-dependent parameter $\kappa_3(T)$ such that

$$H_{c1} = H_c(T)f[\kappa_3(T)]. \tag{VII.9}$$

7

In the limit $\kappa \gg 1$, and hence $\kappa_3 \gg 1$

$$f(\kappa_3) = \ln \kappa_3 / \surd(2)\, \kappa_3. \qquad (VII.10)$$

The detailed calculations predict for $\kappa_3$ a monotonic rise with decreasing temperature and a purity dependence such that $\kappa_3/\kappa = 1\cdot15$ in the pure limit and $\kappa_3/\kappa = 1\cdot53$ in the dirty one. Experimentally it is difficult to pinpoint the value of $H_{c1}$, but the data of Finnemore *et al.* (1966) for niobium and that of Bon Mardion *et al.* (1965) for Pb–Tl alloys appear to give satisfactory agreement.

Many of the calculations described in this section have been confirmed and amplified near $T_c$ by the very detailed treatment of Tewordt and co-workers (Tewordt, 1965a, b; Neumann and Tewordt, 1966b).

## 7.5. Type II superconductors with very high $H_{c2}$

Neither a surface energy nor the size effects discussed in the previous chapter can increase the critical field of a superconductor without limit. Both Clogston (1962) and Chandrasekhar (1962) have pointed out independently that in sufficiently high fields it is no longer correct to assume that the free energy of the *normal* phase is independent of field. With a finite paramagnetic susceptibility $X_p$ (which was ignored in deriving equation II.4), this free energy is, in fact, lowered by an amount $\frac{1}{2}X_p H^2$. Thus, in sufficiently high fields, this alone could already bring about a transition from the superconducting to the normal phase. The limit on the critical field imposed by this mechanism is estimated to be two or three hundred $K$ gauss, and this is consistent with the results of Berlincourt and Hake (1962, 1963). However, calculations by Maki (1964b) which take into account both the spin paramagnetism and the orbital diamagnetism of the electrons predict $H_{c2}$ values *lower* than those actually found in recent measurements (see e.g., Neuringer and Shapira, 1965, 1966a; Kim *et al.*, 1965; Hake, 1967). Werthamer *et al.* (1966) and Maki (1966) have shown that this discrepancy is due to the neglect of the spin-orbit scattering of the electron, which reduces its spin paramagnetism. The measurements of Neuringer and Shapira (1966b) have clearly demonstrated this with a series of similar alloys with varying amounts of spin-orbit scattering.

The calculations taking into account all the spin effects predict temperature variations of $\kappa_1(T)$ and $\kappa_2(T)$ which are only partially confirmed by experiments (Cape, 1966; Hake, 1967). The most striking area of agreement is that both $\kappa_1$ and $\kappa_2$ decrease with decreasing temperature.

Hake (1967) has reported extensive and detailed measurements on very high field alloys which were so carefully prepared as to display *reversible* magnetization. These conclusively demonstrate the existence of *paramagnetic superconductivity*: for a wide range of field in the mixed state below $H_{c2}$, the spin paramagnetism of the specimen exceeds its superconducting diamagnetism, resulting in a net paramagnetic magnetization. At $H_{c2}$ the transition to the normal paramagnetic state is continuous and of second order, as shown also by calorimetric measurements (Barnes and Hake, 1967).

### 7.6. Thin films in a transverse field

Tinkham (1963, 1964) has shown that even for arbitrarily low values of $\kappa_1$ a mixed state vortex structure is energetically favourable for a thin film in a transverse magnetic field. Such a film would become normal in a second-order transition when

$$H_\perp = \sqrt{(2)}\,\kappa_1(T)H_c. \qquad \text{(VII.11)}$$

Tinkham's approximate treatment has been expanded by Pearl (1964), Maki (1965), and Fetter and Hohenberg (1967). The magnetization measurements of Chang *et al.* (1963), the penetration depth and critical field data of Mercereau and Crane (1963), and the results on thickness dependence of the critical field (Guyon *et al.* 1963) support this model. Direct evidence for the vortex structure in the thin films was obtained by Parks and co-workers. Using first thin films so narrow as to contain only a single row of vortices, Parks and Mochel (1964) calculated that the free energy should have a minimum at values if the perpendicular field $H_\perp$ at which a vortex diameter $D = 2\sqrt{(\phi_0/\pi H_\perp)}$ just equals the film width. At $T_c$ this should result in a decrease of the film resistance, and this was indeed observed. However, this might also have been due to induced voltages (Anderson and Dayem, 1964). Parks *et al.* (1964) therefore investigated the critical field behaviour of superconducting bridges, consisting of a narrow film strip linking two larger film areas. At a given temperature, a striking drop in film resistance was found at just that value of external field at which the expected vortex diameter equalled the width of the bridge. Thus the resistance drop appears to correspond with the flow of vortices across the bridge.

Extensive measurements on alloy films of varying purities by Chang and Serin (1966) indicated a mean free path dependence of $\kappa_1(T)$ like that found for bulk type II superconductors (see Section 7.3). In the limit of great purity $\kappa_1(0)/\kappa \to 2$. For impure films the ratio agrees with theory, just as in bulk samples. This has been confirmed by the measurements of Guyon *et al.* (1967).

## 7.7. Surface superconductivity

As mentioned in Chapter V, the boundary condition applicable to the G-L order parameter $\Psi$ is that its derivative vanishes. Saint James and De Gennes (1963) have shown that in an external field parallel to the surface this leads to the persistence of an outer superconducting sheath up to a field

$$H_{c3} = 1 \cdot 695 \sqrt{(2)} \kappa H_c. \tag{VII.12}$$

The thickness of this sheath is of the order of $\xi$. Its existence, explicitly verified by many experiments (see, for example, Hempstead and Kim, 1963; Tomasch and Joseph, 1963; Guyon *et al.*, 1965) explains what has often been a puzzling discrepancy between magnetic and resistive transitions. The sheath is eliminated if the specimen is coated with a layer of normal metal.

The surface sheath exists also in Type I superconductors, but can be detected only if $H_{c3} > H_c$. As $H_{c2} = \sqrt{(2)} \kappa H_c$, it follows that $H_{c3} = 2 \cdot 40 \kappa H_c$, so that $H_{c3} > H_c$ for $\kappa > 0 \cdot 42$. Under this condition, a measurement of $H_{c3}/H_c$ is in fact a way of obtaining $\kappa$ for Type I materials (Strongin *et al.*, 1964; Rosenblum and Cardona, 1964). The parameter relating $H_{c3}(T)$ to $H_c(T)$ is $\kappa_1(T)$ (Maki, 1964a; De Gennes, 1964). Experiments on films of pure lead and dilute mercury alloys (Rosenblum and Cardona, 1964c; Paskin *et al.*, 1965) again show a more rapid temperature variation than predicted by theory. The purity dependence of $\kappa_1(T)$ near $T_c$ has been treated by Ebneth and Tewordt (1965).

Vortex lines can enter the specimen as soon as the perpendicular component of the external field reaches even a very small value and as a result the value of $H_{c3}$ decreases rapidly with small deviations from a parallel orientation. The angular dependence has been calculated by Tinkham (1964) and Saint James (1965) and measured by Burger *et al.* (1965) and others.

That the thickness of the surface sheath is of the order of the coherence length has been demonstrated by the thermal conductivity measurements of Seidel and Meissner (1966). Mochel and Parks (1966) have shown that the variation of thermal conductivity with field is in excellent agreement with the theory of Caroli and Cyrot (1965).

Considerable theoretical and experimental work has been done on the current-carrying capacity of the surface sheath (see e.g., Fink, 1966).

*Note added in proof*: The latest work of Fiory and Serin (*Phys. Rev.Lett.* **21,** 359, 1968) shows rather conclusively that the Hall angle does remain constant, at least in pure niobium. (See page 84.)

# CHAPTER VIII

# The Isotope Effect

## 8.1. Discovery and theoretical considerations

The various phenomenological treatments based on the empirical characteristics of a superconductor provide an astonishingly complete macroscopic description of the superconducting phase. However, they do not give any clear indications as to the microscopic nature of the phenomenon.

One of the first such clues arose through the simultaneous and independent discovery, in 1950, by Maxwell, and by Reynolds *et al.*, that the critical temperature of mercury isotopes depends on the isotopic mass by the relation

$$T_c M^a = \text{constant}, \qquad \text{(VIII.1)}$$

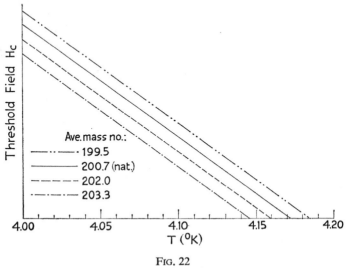

FIG. 22

where $M$ is the isotopic mass and $a \approx \frac{1}{2}$. This is illustrated in Figure 22, showing the variation of threshold field near $T_c$ for different isotopes. The effect has since also been established in a number of other elements. The following table contains the most reliable experimental values of the exponent $a$, together with quoted probable errors.

| Element | $a$ | References |
|---|---|---|
| Cd | $0.51 \pm 0.10$ | Olsen, 1963 |
| Hg | $0.504$ | Reynolds *et al.*, 1951 |
| Mo | $0.33$ | Matthias *et al.*, 1963 |
|  | $0.33 \pm 0.05$ | Bucher *et al.*, 1965 |
| Os | $0.21$ | Hein and Gibson, 1963 |
|  | $0.20 \pm 0.05$ | Bucher *et al.*, 1965 |
| Pb | $0.461 \pm 0.025$ | Shaw *et al.*, 1961 |
|  | $0.501 \pm 0.013$ | Hake *et al.*, 1958 |
| Re | $0.4$ | Maxwell and Strongin, 1964 |
|  | $0.39 \pm 0.01$ | Bucher *et al.*, 1965 |
| Ru | $< 0.1$ | Geballe *et al.*, 1961 |
|  | $< 0.05$ | Finnemore and Mapother, 1962 |
|  | $0.0 \pm 0.10$ | Bucher *et al.*, 1965 |
|  | $0.0 \pm 0.15$ | Gibson and Hein, 1966 |
| Sn | $0.505 \pm 0.019$ | Maxwell, 1952a |
|  | $0.46 \pm 0.02$ | Serin *et al.*, 1952 |
|  | $0.462 \pm 0.014$ | Lock *et al.*, 1951 |
| Tl | $0.50 \pm 0.05$ | Maxwell, 1952b |
|  | $0.62 \pm 0.1$ | Alekseevskii, 1953 |
| Zn | $0.45$ | Geballe and Matthias, 1964 |
|  | $0.30$ | Fassnacht and Dillinger, 1966 |
| Zr | $0.0 \pm 0.05$ | Bucher *et al.*, 1965 |

In all the non-transition metals, with the exception of molybdenum, the results are consistent with $a = \frac{1}{2}$. However, small mass differences and the possibility of impurity and strain effects limit the experimental reliability, as is made evident by the variations between different measurements on the same element. Thus one cannot rule out deviations from the ideal value of $a = \frac{1}{2}$ which may be as high as 20 per cent in some cases. In view of recent theoretical work to be discussed in Chapter XI, it is significant that the trend of the published deviations from $a = \frac{1}{2}$ is toward lower values. The situation in the transition metals ruthenium and osmium, however, appears to be different. This will be further discussed in Section 11.4.

The inference to be drawn from the dependence of $T_c$ on the isotopic mass is startling. A relation between the onset of superconductivity, which is quite certainly an electronic process, and the isotopic mass, which affects only the phonon spectrum of the lattice, must mean that superconductivity is very largely due to a strong interaction between the electrons and the lattice. Thus the discovery of the isotope effect clearly pointed out the direction in which a microscopic explanation of the phenomenon had to be sought. In fact, Fröhlich (1950) had independently suggested just such a mechanism without knowing of the experimental work. However, it took several more years until the subtle nature of the pertinent electron–lattice interaction was recognized and a valid microscopic theory began to be developed.

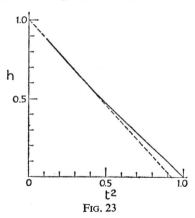

FIG. 23

## 8.2. Precise threshold field measurements

The variation of critical temperature with isotopic mass was established by measuring the critical field $H_c$ as a function of temperature, and then extrapolating this to zero field. Magnetic measurements of course make use of the perfect diamagnetism of a superconductor, and can be made in one of two ways: either the change in flux through the sample at the transition induces an e.m.f. in a pick-up coil which is connected to a suitable galvanometer, or the changing susceptibility of the sample is reflected in the change of the mutual inductance of coaxial coils of which the sample forms part of the core. Either of these methods can be applied with great accuracy in spite of simple apparatus, and has the further advantage of measuring a bulk property virtually unaffected by the possible presence of small regions with different superconducting characteristics. By providing a misleading short-circuiting path, such minor flaws can lead to very

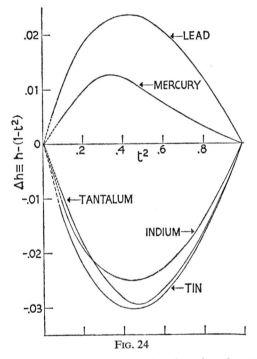

FIG. 24

erratic results when $T_c$ is measured by observing the variation oj electrical resistance.

The careful determination of critical field curves which arose as almost a by-product of the work on the isotope effect established a number of interesting characteristics. Figure 23 shows the variation of the reduced critical field $h \equiv H_c/H_0$ as function of $t^2 \equiv T^2/T_c^2$ for a number of tin isotopes measured by Serin *et al.* (1952). It is evident, as was indicated earlier, that equation I.2a is only an approximation, and that a better representation for $h$ is a polynomial

$$h(t) = 1 - \sum_{n=2}^{N} a_n t^n.$$  (VIII.2)

A polynomial which fits the data for all tin isotopes as found by Lock *et al.* (1951) to within one-half of a per cent is

$$h = 1 - 1 \cdot 0720 t^2 - 0 \cdot 0944 t^4 + 0 \cdot 3325 t^6 - 0 \cdot 1660 t^8.$$  (VIII.3)

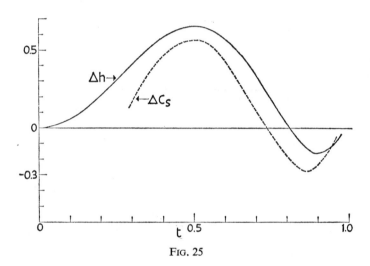

Fig. 25

All measurements to date have indicated that to within the available precision all isotopes of a given element follow the same critical field polynomial. One also finds that $H_0$ has the same mass dependence as $T_c$. This means that the superconducting condensation energy $H_0^2/8\pi$ varies proportionally to the isotopic mass, and also that, as shown by equations II.15 or II.16, the value of $\gamma$ is independent of isotopic mass.

It has also been found that in going from one element to another, the reduced threshold field curves show small but definite variations.

For all elements there are deviations from a strictly parabolic variation, generally by a similar small amount in one direction, but in the case of lead and mercury by an amount in the opposite direction. Figure 24 shows these deviations as a function of reduced temperature. It is important to emphasize the smallness of these deviations, so as not to allow them to obscure the basic similarity of the superconducting behaviour of all elements in terms of reduced co-ordinates. This not only sanctions the continuing discussion of superconductivity in general terms with only occasional references to specific elements, but also allows one to look for a microscopic explanation of superconductivity, which in first approximation need not concern itself with the distinctive characteristics of individual elements, but only takes account of general and common features. The deviations of the measured threshold fields from a simple parabolic variation must be, according to the thermodynamic treatment developed in Chapter II, correlated with the empirical deviations of the specific heat from the corresponding change as the cube of the temperature. Serin (1955) showed this strikingly by plotting both these deviations on the same graph, using the best available data for tin. This is shown in Figure 25. Mapother (1959) has since established the correlation between the experimental non-parabolic threshold fields and the exponential variation of the specific heat.

# CHAPTER IX

# Thermal Conductivity

## 9.1. Low temperature thermal conductivity

In normal metals, heat is carried both by the conduction electrons and by the quantized lattice vibrations, the phonons. The total thermal conductivity consists of the sum of these two contributions:

$$k_n = k_{en} + k_{gn}, \tag{IX.1}$$

where $e$ and $g$ denote the electrons and the lattice, respectively. The electronic conductivity is limited by two scattering mechanisms: the phonons and the lattice imperfections, and one can write at $T < \Theta$:

$$1/k_{en} = aT^2 + \rho_0/LT. \tag{IX.2}$$

The first of the terms on the right gives the resistivity due to the electron scattering by phonons, and predominates at higher temperatures; the second that due to scattering by imperfections, which becomes important below the temperature at which $k_{en}$ has a maximum:

$$T_{\max}^3 = \rho_0/2aL. \tag{IX.3}$$

In these equations $\rho_0$ is the residual electrical resistivity, $L$ the Lorentz number $(2 \cdot 44 \times 10^{-8}$ watt-ohm/deg$^2)$ and $a$ is a constant of the material which is inversely proportional to $\Theta^2$. Note that for a given material the addition of impurities increases $\rho_0$ and thus raises $T_{\max}$. In pure metals and dilute alloys, $k_{en} \gg k_{gn}$; it is only in metals containing as much as several per cent impurities that the two contributions are of the same order of magnitude.

The two-fluid model allows one to predict qualitatively what happens to the thermal conductivity of a metal when it becomes superconducting (see Mendelssohn, 1955; Klemens, 1956). The condensed 'superconducting' electrons cannot carry thermal energy nor can they scatter phonons. With decreasing temperature their number increases, and that of the 'normal' electrons correspondingly decreases, which will result in a rapid decrease of the electronic heat conduction. At the same time the conduction by phonons will be enhanced, as these are no longer scattered as much by electrons.

In pure specimens, the decrease in $k_{es}$ will usually exceed any gain in $k_{gs}$, and the total conductivity in the superconducting phase will then be much smaller than in the normal phase. This is illustrated, for example, by the results of Hulm (1950) on pure Hg shown in

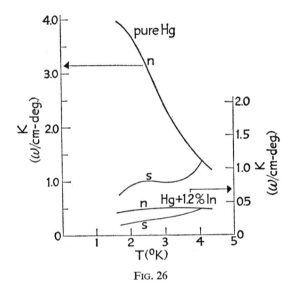

Fig. 26

Figure 26. There exist, however, pure materials in which the normal conductivity is not very high but which are very free of grain boundaries and other lattice defects. In the superconducting phase of such substances at very low reduced temperatures the phonons are then hardly scattered by anything except the specimen boundaries, resulting in a large value of $k_{gs}$. This has been observed, for example, by Calverley *et al.* (1961) in tantalum and niobium.

Suppressing the electronic conduction in the normal phase by adding impurities decreases the effect of condensing electrons out of the thermal circuit. For moderately impure specimens the super-conducting conductivity will then not be very different from the corresponding normal one. This is shown, for example, by the results of Hulm (1950) on a Hg–In alloy, also displayed in Figure 26. The results of Lindenfeld (1961) on lead alloys shown in Figure 27 indicates what happens with increasing inpurity content: as the phonon contribution to the normal conductivity becomes more appreciable, the

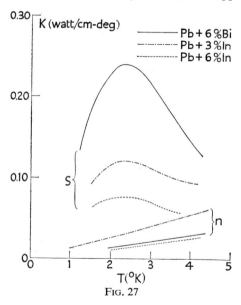

Fig. 27

gain in $k_{gs}$ increasingly outweighs the decrease in $k_{es}$, and the conductivity in the superconducting phase becomes much larger than that in the normal one.

### 9.2. Electronic conduction

If the effect on thermal conductivity by the superconducting transition is indeed due to the disappearance of electrons from the conduction process, then one should be able to write IX.2 for a superconductor as

$$1/k_{es} = x(\mathscr{W})aT^2 + y(\mathscr{W})\rho_0/LT, \qquad (IX.4)$$

where $x(\mathscr{W})$ and $y(\mathscr{W})$ are functions only of the order parameter $\mathscr{W}$ which indicates the fraction of condensed electrons. Equation II.25 shows that $\mathscr{W}$ is a function only of $t \equiv T/T_c$, so that one can write instead

$$1/k_{es} = aT^2/g(t) + \rho_0 LT/f(t). \qquad (IX.4)$$

The equation has been written in this form to agree with the nomenclature introduced by Hulm (1950). He pointed out that if one chooses a sample in which the electronic heat conduction is predominantly limited by one or the other of the two scattering mechanisms, the measured ratio $k_{es}/k_{en}$ then equals the appropriate ratio function $g(t)$ or $f(t)$. To a first approximation, at least, these functions should be universal functions for all superconductors and be related to the microscopic nature of the phenomenon.

For a specimen for which $T_{\max} < T_c$, as is the case for reasonably pure Hg and Pb, and for extremely pure Sn and In, the heat conduction just below $T_c$ is by electrons limited by phonon scattering. For such samples

$$k_{es}/k_{en} \approx g(t). \qquad (IX.5)$$

All pertinent measurements show the same qualitative features: $g(t)$ at $t = 1$ breaks away sharply from unity with a discontinuous slope, and decreases as a power of $t$ which is about 2 for Sn and In (Jones and Toxen, 1960: Guenault, 1960), and 4 to 5 for Pb and Hg (Watson and Graham, 1963; see also Klemens, 1956). Calculations by Kadanoff and Martin (1961), by Kresin (1959) and by Tewordt (1962, 1963a) appear to explain the experimental results for Sn and In. For Hg and Pb it is necessary to take into account so-called strong coupling effects (see Section 11.8), (Ambegaokar and Tewordt, 1964; Ambegaokar and Woo, 1965).

For specimens for which $T_{\max} \geqslant T_c$, the electronic conduction in the superconducting phase is at all temperatures limited by impurity scattering, so that for these

$$k_{es}/k_{en} \approx f(t). \qquad (IX.6)$$

Several investigations (see Klemens, 1956) have shown that at $t = 1$ $f(t)$ approaches unity smoothly with a continuous slope, and that at lower temperatures it decreases more slowly than $g(t)$. The results are in reasonable agreement with expressions for $f(t)$ derived by Bardeen *et al.* (BRT, 1959) and by Geilikman and Kresin (1959) on the basis of the BCS theory. The gradual change from a phonon-scattered to an impurity-scattered electronic conduction in the same material of increasing impurity is particularly well illustrated by the recent results of Guénault (1960) on a series of monocrystalline tin specimens.

When thermal conductivity measurements on superconductors are extended to small values of $t$, as was first done by Heer and Daunt (1949) and later by Goodman (1953), $f(t)$ is found to decrease very rapidly. Goodman pointed out that this could be represented by an equation of the form

$$f(t) = a\exp(-b/t), \tag{IX.7}$$

and suggested that this implied the existence of an energy gap between the ground state and the lowest excited state available to the assembly of superconducting electrons.

This conclusion can be inferred from thermal conductivity results in the following manner. Simple transport theory shows that

$$k_e = (1/3)\,lv_0\,C_e, \tag{IX.8}$$

where $l$ is the mean free path, $v_0$ the average velocity, and $C_e$ the specific heat of the electrons. Assuming that $v_0$, the Fermi velocity in the normal metal, remains the same for the uncondensed 'normal' electrons in the superconducting phase, and that in both phases the mean free paths (which may differ in magnitude) vary only slowly with temperature, then the temperature variation of $k_{es}/k_{en}$ must be due entirely to that of the specific heats. In other words

$$f(t) \approx k_{es}/k_{en} \approx C_{es}/C_{en}. \tag{IX.9}$$

$C_{en}$ is known to vary linearly with temperature, so that IX.7 implies that

$$C_{es} = a'T_c\,t\exp(-b/t). \tag{IX.10}$$

That such a temperature variation of the specific heat corresponds to an energy gap in the electronic spectrum can be shown as follows: If a gap of width $2\epsilon$ lies below the lowest available excited state, the number of thermally excited electrons will be proportional to

$\exp(-2\epsilon/2k_BT)$, where $k_B$ is the Boltzmann constant, and the factor 2 arises because every excitation creates two independent particles, an electron and a hole. Thus the free energy of the superconducting phase is equal to the condensation energy per particle multiplied by the exponential factor, which remains unchanged, through two differentiations with respect to temperature, to appear in the specific heat. The parameter $b$ in IX.10 is thus seen to equal $2\epsilon/2k_BT_c$.

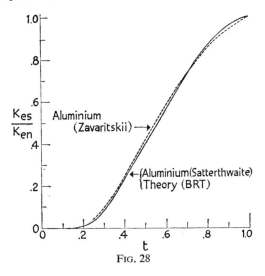

FIG. 28

According to the microscopic theory to be discussed in Chapter XI, the energy gap is a function of temperature. The parameter $b$ can therefore be written as

$$b = \frac{\epsilon(T)}{k_BT_c} = \frac{\epsilon(0)}{k_BT_c} \times \frac{\epsilon(T)}{\epsilon(0)}$$

where $\epsilon(0)$ is the gap value at $0°K$. The detailed dependence of $k_{es}/k_{en}$ on $b$ has been calculated by BRT, and the function $\epsilon(T)/\epsilon(0)$, calculated from the BCS theory, has been tabulated by Mühlschlegel (1959). Measurements of $k_{es}/k_{en}$ can thus be used to infer the value of $\epsilon(0)/k_BT_c$.

8

The appropriate temperature dependence of $k_{es}/k_{en}$ has been observed in a number of metals. The results for aluminium by Satterthwaite (1960) and by Zavaritskii (1958a) are shown in Figure 28, together with a theoretical curve calculated with a gap equal to $3 \cdot 50 \, k_B T_c$. The agreement is somewhat deceptive, since there is good evidence that the gap width for aluminium is only $3 \cdot 40 \, k_B T_c$. From an observed anisotropy in the temperature dependence of $k_{es}$ at very low temperatures Zavaritskii (1959, 1960a, b) has been able to infer a corresponding anisotropy in the width of the energy gap in the spectrum of the superconducting electrons in the case of cadmium, tin, gallium, and zinc. To the last he could apply theoretical expressions due to Khalatnikov (1959), from which he deduced a gap anisotropy of about 30 per cent. A similar result holds for cadmium. The measurements of Zavaritskii also show that the gap anisotropy can have different forms: in the case of gallium the value of the gap can be approximated by an ellipsoid compressed along the axis of rotation; for zinc and cadmium this ellipsoid is stretched out along the axis of rotation.

In cases where the energy gap is a function of the magnetic field, measurements of $k_{es}/k_{en}$ can be used to infer this field dependence. This technique has been used by Morris and Tinkham (1961) for thin films (see Section 7.3), and by Dubeck *et al.* (1962, 1964) for type II superconductors in the mixed state.

The measurements in the mixed state show a striking dependence of the thermal conductivity on magnetic field. Below $H_{c1}$, most of the heat is conducted by the lattice, and the component drops very abruptly when $H \geq H_{c1}$, probably due to the scattering of phonons from the vortex lines. As the field further increases toward $H_{c2}$, the average energy gap decreases, there are more normal electrons, and the electron component of the conductivity increases rapidly. In bulk samples the interplay of these two mechanisms on the total conductivity leads to a rapid dip followed by a rise. In films, in which the lattice component is small even below $H_{c1}$, only the rise due to the electronic component is observed (Mochel and Parks, 1966; Parks *et al.*, 1967, Smith and Ginsberg, 1967). Anisotropy effects observed both in bulk and thin film measurements have been further studied by Lindenfeld and McConnell (1967).

Comparison with the microscopic theory of Caroli and Cyrot (1965) is difficult because it has been derived only in the dirty limit with $\kappa \gg 1$. However, agreement with both the film measurements and those on bulk specimens (Lindenfeld *et al.*, 1966) is very good.

## 9.3. Lattice conduction

Far below $T_c$ the fraction of 'normal' electrons becomes so small as to make $k_{es} \ll k_{gs}$. At the very lowest temperatures, the phonons are primarily scattered by crystal boundaries in a manner which is the same in the superconducting as in the normal phase. The characteristic $T^3$ dependence in this limit (Casimir, 1938) has been well established experimentally (Mendelssohn and Renton, 1955; Graham, 1958).

In the normal state there occurs at these temperatures still appreciable heat conduction by electrons, limited only by impurity scattering and varying linearly with temperature (see equation IX.2). Thus in this range

$$k_n/k_s = aT^{-2}, \tag{IX.11}$$

where $a$ is a constant of the material which can have values as high as several hundred. For example, a suitable lead wire can have $k_n/k_s \approx 10^5$ at $0.1°$K. A number of authors (see Mendelssohn, 1955) suggested using such wires in ultra-low temperature experiments as thermal switches which would be 'open', i.e. non-conducting, in the superconducting phase, and 'closed' when the superconductivity is quenched by means of a suitable magnetic field. Such heat switches are now widely used (see, for instance, Reese and Steyert, 1962).

At somewhat higher temperatures, at which the phonons begin to be scattered by the 'normal' electrons even in the superconducting phase, there is necessarily a concurrent rise of the electronic conduction. Experimentally it is very difficult to separate the conduction mechanisms. Where this has been possible (Conolly and Mendelssohn, 1962; Lindenfeld and Rohrer, 1965) the results have been consistent with the pertinent calculations by BRT and by Geilikman and Kresin (1958, 1959).

### 9.4. The thermal conductivity in the intermediate state

A number of experiments, in particular those of Mendelssohn and co-workers (Mendelssohn and Pontius, 1937; Mendelssohn and Olsen, 1950; Mendelssohn and Shiffman, 1959), have shown that the thermal conductivity of a superconductor in the intermediate state generally does not change linearly from its value in the one phase to its value in the other when at a given temperature the external field is varied. Instead there appears an extra thermal resistance, which in some cases can be very large, and which is attributed to the scattering of the predominant heat carriers (electrons or phonons) at the boundaries between the superconducting and normal laminae which make up the intermediate state. For materials in which phonon conduction dominates this has been analyzed by Cornish and Olsen (1953) and by Laredo and Pippard (1955). Strässler and Wyder (1963) have developed a treatment for very pure specimens in which the conduction is mostly by electrons. Andreev (1964) has further developed the reflection mechanism for this case, and finds a result independent of electron mean free path, in agreement with the measurements of Zavaritskii (1960a).

Experiments on the thermal conductivity in the intermediate state thus yield strong confirmation that the laminar structure which is observed at the surface of a specimen (see Section 6.2) actually persists throughout a bulk sample.

# CHAPTER X

# The Energy Gap

## 10.1. Introduction

Ever since the initial discovery of superconductivity it had been known but barely noted that the striking electromagnetic behaviour of a superconductor at low frequencies is not accompanied by any corresponding changes in its optical properties: there is no visible change at $T_c$, although the reflectivity of a metal at any frequency is related to its conductivity at that frequency. Thus at the very high optical frequencies the resistance of a superconductor is a constant, independent of temperature, and equal to that of the normal metal. At about the time of the discovery of the isotope effect steadily improving high frequency techniques had shown that at $0°K$ the normal resistance persisted down to frequencies of the order of $10^{13}$ c/sec, but that it remained zero up to frequencies of the order $10^{10}$ c/sec. In 1952 already Shoenberg ([1], p. 202) concluded from this that at some frequency between these two limits '... quantum processes set in which could raise electrons from the condensed to the uncondensed state and thus cause energy absorption'.

As shown in the previous chapter, Goodman (1953) very shortly after this inferred from his thermal conductivity results the existence of an energy gap in the single electron energy spectrum. A similar conclusion had been deduced a few years earlier by Daunt and Mendelssohn (1946) from the absence of any Thomson heat in the superconducting state. This indicated to them that the superconducting electrons remain effectively at $0°K$ up to $T = T_c$ by being in low-lying energy states separated from all excited states by an energy gap of the order $\kappa_B T_c$. Ginzburg (1944) had even earlier considered the possibility of a gap in the superconducting spectrum and of resulting quantum absorption.

In the years which followed, the existence of such a gap was firmly established by a large number of experiments, and this, together with the electron–phonon interaction indicated by the isotope effect, provided the keystones of a microscopic theory. This chapter will describe a few experiments which indicate the energy gap most clearly and directly. The subject has been reviewed by Biondi *et al.* (1958) and by Douglass and Falicov (1964).

In anticipation of a later discussion (see Section 11) it is useful to stress here that an energy gap is said to exist if there is a complete absence of low-lying energy states over a finite range of energies. This indeed is the prevalent situation in superconductivity. However, in certain cases superconductivity can also occur if there is merely a very low density of states (rather than a total absence) over a certain energy range. This has become known as 'gapless' superconductivity.

## 10.2. The specific heat

After the resurgence of interest in specific heat measurements as a result of the suggestive results of precise threshold field measurements, of Goodman's thermal conductivity results, and of the first clear experimental verification of a deviation from a $T^3$ law by Brown *et al.* (1953) on niobium, there have been in recent years a number of measurements which clearly indicate the exponential variation of $C_{es}$ corresponding to an energy gap. The first of these were the results of Corak *et al.* (1954) on vanadium and by Corak and Satterthwaite (1954) on tin; and since then the exponential variation of $C_{es}$ has been established in a number of elements. The appropriate column in Table III lists the energy gap values of these elements deduced from the specific heat measurements. Note that in units of $k_B T_c$ these gaps are of very similar size for widely varying superconductors. This again bears out the basic similarity of all superconductors in terms of reduced co-ordinates.

It is perhaps useful to consider briefly the difficulty of obtaining good values for $C_{es}$. What is measured, of course, in both the superconducting and in the normal phase, is the total specific heat. It is then necessary to separate the electronic from the lattice contribution in the normal phase in order to be able to subtract the latter from the total specific heat in the superconducting phase. Unfortunately, even at low temperatures, $C_{gn}$ is small compared to $C_{en}$ only for metals with large Debye temperatures. These are just the hard, high-melting point metals which are difficult to obtain with high purity, without which superconducting measurements are misleading. The softer and lower melting point metals, on the other hand, have a very unfavourable ratio of electronic to lattice specific heat.

Measurements by Goodman (1957, 1958), Zavaritskii (1958b), and Phillips (1959) on aluminium have shown at very low temperatures

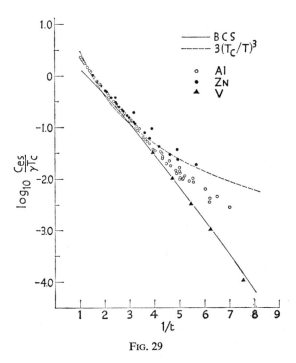

Fig. 29

$(t < 0 \cdot 2)$ a deviation of $C_{es}$ from a simple exponential law (Boorse, 1959). Cooper (1959) has pointed out that this can be a consequence of anisotropy in the energy gap. At the lowest temperatures most electron excitations would be expected to occur across the narrower portions of the gap, and this would be reflected in an upward curvature of $C_{es}$ when plotted semi-logarithmically against $1/t$. Figure 29 compares a number of measurements which show this curvature with the exponential law expected from the BCS theory. According to a theory of Anderson (1959) (see Section 12.2) the gap anisotropy of an element diminishes with the addition of impurities. Indeed Geiser and Goodman (1963) have found in aluminium specimens of different purity that the deviation of $C_{es}$ from an exponential form decreases with increasing impurity, and very pure vanadium shows the upward deviation from an exponential law (Radebaugh and Keesom, 1966a).

Recent very careful measurements of specific heats at very low temperatures ($1/t > 5$) have demonstrated the existence of two quite distinct slopes, indicating the presence of two different energy gaps. This has been found for lead (Vander Hoeven and Keesom, 1965), niobium, tantalum, and vanadium (Shen *et al.*, 1965; Radebaugh and Keesom, 1966a). Either or both of these gaps – which may be due to the overlapping of $s$ and $d$ bands at the Fermi surface (Suhl *et al.*, 1959b; Sung and Shen, 1965) – may be anisotropic.

### 10.3. Electromagnetic absorption in the far infrared

The magnitude of the energy gap $2\epsilon(0)$ can be characterized by a frequency $\nu_g$ such that $h\nu_g = 2\epsilon(0)$. It is at this frequency that one would expect the change from the characteristically superconducting response to low frequency radiation, to the normal resistance maintained at high frequencies. Unfortunately, the frequencies corresponding to gap widths inferred from the specific heat measurements are $10^{11}$–$10^{12}$ c-sec$^{-1}$, which is an experimentally awkward range at the upper limit of klystron-excited frequencies, yet very low for mercury arc ones. Only recently have Tinkham and collaborators developed the techniques needed to detect the very low radiation intensities available in this far infrared region.

The measurements of the transmission of such radiation through superconducting films will be discussed in a later section. Richards and Tinkham (1960), Richards (1961), and Ginsberg and Leslie (1962) have observed directly the absorption edge at the gap frequency in bulk superconductors. Radiation from a quartz mercury arc infrared monochromator was fed by means of a light pipe into a cavity made of the superconducting material under investigation. The cavity contained a carbon resistance bolometer, and was shaped so that the incident radiation would make many reflections before striking this detector. For frequencies lower than $\nu_g$, the superconducting walls of the cavity do not absorb, and much radiation reaches the bolometer. At $\nu_g$, absorption by the walls sets in, and the signal from the bolo-

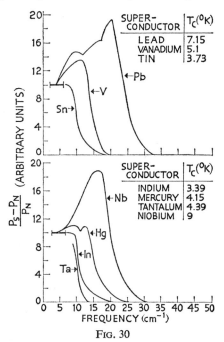

| SUPER-CONDUCTOR | $T_c(^{\circ}K)$ |
|---|---|
| LEAD | 7.15 |
| VANADIUM | 5.1 |
| TIN | 3.73 |

| SUPER-CONDUCTOR | $T_c(^{\circ}K)$ |
|---|---|
| INDIUM | 3.39 |
| MERCURY | 4.15 |
| TANTALUM | 4.39 |
| NIOBIUM | 9 |

FIG. 30

meter decreases sharply. Figure 30 shows normalized curves of the fractional change in the power absorbed by the bolometer, in arbitrary units, plotted against frequency for all the metals investigated by Tinkham and Richards. The gap values obtained are listed in Table III. Early measurements indicated the presence of a so called *precursor* structure in the absorption edges, found also in infrared transmission (Ginsberg *et al.*, 1959), which appeared to persist even in very impure specimens (Leslie and Ginsberg, 1964). However,

# 112 *Superconductivity*

very recent measurements have shown the precursor effect to be spurious and probably due to the leakage of high-frequency radiation (see e.g., Palmer and Tinkham, 1966). Prior attempts to explain

TABLE III

| Element | Energy gap $(2\epsilon(0)/k_B T_c)$ | | | | | |
|---|---|---|---|---|---|---|
| | A | B | C | D | E | F |
| Aluminium | ... | ... | 3·16 | 2·9 | 3·37[b], 3·43[c] 3·3[d] | 3·5 |
| Cadmium | ... | ... | ... | ... | ... | 3·3 |
| Gallium | ... | ... | ... | ... | ... | 3·5 |
| Indium | 4·1 | 3·9 | ... | 3·9 | 3·63[a], 3·45[b] | 3·5 |
| Lanthanum | 2·85 | ... | ... | ... | 3·2[i] | 3·7 |
| Lead | 4·14 | 4·0 | ... | ... | 4·33[a], 4·26[b] 4·18[d] | 3·9 |
| Mercury$\alpha$ | 4·6 | ... | ... | ... | 4·6[k] | 3·7 |
| Niobium | 2·8 | ... | ... | 4·4 | 3·84[e], 3·6[f], 3·59[g] | 3·7 |
| Rhenium | ... | ... | ... | ... | ... | 3·3 |
| Ruthenium | ... | ... | ... | ... | ... | 3·1 |
| Tantalum | $\leqslant 3·0$ | ... | ... | 3·6 | 3·60[e], 3·5[f], 3·65[h] | 3·6 |
| Thallium | ... | ... | ... | 3·2 | 3·57 ± 0·05[l] 3·9[m] | 2·8 |
| Thorium | ... | ... | ... | ... | ... | 3·5 |
| Tin | 3·6 | 3·3 | 3·5 | 3·6 | 3·46[a], 3·47[b] 3·65[d] | 3·6 |
| Vanadium | 3·4 | ... | ... | 3·6 | 3·4[f] | 3·6 |
| Zinc | ... | ... | ... | 2·5 | 3·2 ± 0·1[n] | 3·4 |

A—from infrared absorption (lead: Ginsberg and Leslie, 1962; lanthanum: Leslie *et al.*, 1964; all others Richards and Tinkham, 1960).

B—from infrared transmission (Ginsberg and Tinkham, 1960).

C—from microwave absorption (aluminium: Biondi and Garfunkel, 1959; tin: Biondi *et al.*, 1957).

D—by fitting specific heat data to exponential (Goodman, 1959).

E—from tunneling ([a]Giaever and Megerle, 1961; [b]Zavaritskii, 1961; [c]Douglass, 1962; [d]Douglass and Meservey, 1964a; [e]Townsend and Sutton, 1962; [f]Giaever, 1962; [g]Sherrill and Edwards, 1962; [h]Dietrich, 1962; [i]Hauser, 1966; [k]Bermon and Ginsberg, 1964; [l]Taylor and Burstein, 1963; [m]Rowell and McMillan, 1966; [n]Donaldson, 1966).

F—calculated from XI.32 (Goodman, 1959).

the effect as being due to states of collective excitations lying in the gap (Tsuneto, 1960; Larkin, 1964) had been unable to account for its apparent insensitivity to impurities (Maki and Tsuneto, 1962; Fulde and Strassler, 1965).

Richards (1961) has reported measurements on single crystals of pure tin and of tin containing 0·1 atomic per cent indium. His results show that the position of the absorption edge varies with crystal orientation, which clearly indicates the anisotropy of the gap. Furthermore this anisotropy decreases with increasing impurity, which strongly supports Anderson's suggestion (1959) that the anisotropy becomes smoothed out in impure samples. The absorption edges observed by Richards have a structure which, unlike that seen in Pb and in Hg, occurs for frequencies greater than $\nu_g$. These postcursor peaks do not seem to change with impurity, and have not yet found an explanation.

### 10.4. Microwave absorption

Although the resistivity of a superconductor vanishes at $0°K$ for frequencies up to $\nu_g$, there is a finite resistance at higher temperatures even at lower frequencies (H. London, 1940). One can understand this from a simple two-fluid picture, according to which at any finite temperature a fraction of the electrons remains 'normal'. H. London pointed out that in the presence of an alternating electric field these electrons absorb energy as they would in a normal metal, and that such a field is needed to sustain an alternating current even in a super conductor because of the inertia of the superconducting electrons.

Into a normal metal an alternating field penetrates to a skin depth $\delta$, which leads to anomalous results if the mean free path $l \gg \delta$, as is the case at high frequencies and low temperatures (see p. 44). In the superconducting phase, the theory of the anomalous skin effect still applies in principle, but has to be modified both because for high frequencies the superconducting penetration depth $\lambda$ is much smaller than the skin depth $\delta$ (except very near $T_c$) and decreases very rapidly with decreasing temperature, and because the number of 'normal' electrons also drops sharply below $T_c$. Both of these lead to a reduction of the resistance in the superconducting phase as compared to that in the normal one: the ratio of the resistances decreases rapidly below $T_c$, changes more gradually at lower temperatures where both $\lambda$ and the order parameter are fairly constant, and finally vanishes at 0°K where there are no more 'normal' electrons.

Unpublished calculations of the variation of $R_s/R_n$ with temperature and with frequency have been made by Serber and by Holstein on the basis of the Reuter-Sondheimer equations, the London theory, and the two-fluid model. Typical results are the solid curve labelled $0.65k_BT_c$ and the dashed one labelled $2.37k_BT_c$ in Figure 31. With

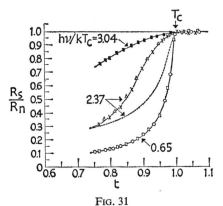

Fig. 31

frequencies up to $8 \times 10^{10}$ c/sec there is general experimental agreement with these calculations, as shown, for example, by the recent results of Khaikin (1958) on cadmium and of Kaplan *et al.* (1959) on tin. Their temperature dependence for a given frequency can be represented by an empirical function, suggested by Pippard (1948):

$$\phi(t) = t^4(1-t^2)(1-t^4)^{-2}. \tag{X.1}$$

The frequency dependence is as $\nu^{4/3}$ at low frequencies, tending toward a constant value at higher frequencies.

However, surface impedance measurements at frequencies considerably higher than $8 \times 10^{10}$ c/sec show appreciable deviations from the predictions of the simple two-fluid model. Figure 31 shows the ratio $R_s/R_n$ as a function of reduced temperature for aluminium as measured at three microwave frequencies by Biondi *et al.* (1957). The frequencies are given in units of $k_B T_c/h$. For 0·65, the results agree well with the temperature variation calculated without regard to an energy gap. For 2·37, however, such calculations would give the dashed curve, and it is evident that for $t > 0·7$, the measured ratio considerably exceeds the predicted one. The same is true for $hv = 3·04 k_B T_c$, except that in this case the deviation already begins at lower $t$.

Clearly an additional absorption mechanism occurs for these frequencies, and of course this is due to the boosting of condensed electrons across the energy gap. If this gap had a constant width at all $t < 1$, the appearance of this extra absorption would depend only on the frequency. Its temperature dependence, however, clearly shows that the energy gap varies with temperature, tending toward zero as $t \to 1$. As a result, photons of energy $2·37 k_B T_c$, for example, are not sufficient to bridge the gap at $t = 0$, but become effective at that temperature at which the gap has shrunk to a width of $2·37 k_B T_c$. A series of measurements of the resistance ratio as a function both of frequency and of temperature thus serves to map out the temperature variation of the gap of any given superconductor. Biondi and Garfunkel (1959) have obtained values of the resistance ratio by measuring calorimetrically the amount of energy absorbed by an aluminium wave guide, over a range of frequencies ranging from $0·65 k_B T_c$ ($1·5 \times 10^{10}$ c/sec) to $3·91 k_B T_c$ ($10 \times 10^{10}$ c/sec) at temperatures down to 0·35°K. The accuracy of the measurements was such that the absorption of $10^{-9}$ watt could be detected. Their results give a temperature variation of the gap which is in close agreement with the predictions of the BCS theory.

Mattis and Bardeen (1958) and Abrikosov *et al.* (1958) have developed a theory of the anomalous skin effect in superconductors on the basis of the BCS theory. Miller (1960) used the work of the former to calculate the surface impedance for many different frequencies and temperatures. The close agreement between his results and the measurements of Biondi and Garfunkel is shown in Figure 32, in which points calculated by Miller are superimposed on smooth curves representing the empirical values. The theoretical treatments are equally successful in the lower frequency range in which there are no gap effects.

Further absorption measurements on aluminium by Biondi *et al.* (1964) clearly indicated the existence of two distinct energy gaps, as had already been found for some transition metals by specific heat experiments (see Section 10.2). Both of these two gaps have an anisotropy which disappears with decreasing mean free path (Biondi *et al.*, 1965).

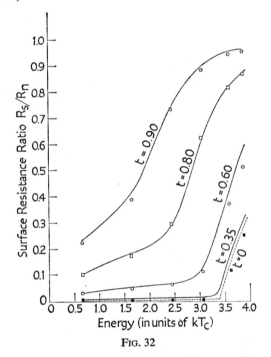

Fig. 32

In the presence of a magnetic field the absorption of microwave radiation is strongly enhanced (Budzinski and Garfunkel, 1966). Pincus (1967) has suggested that this may be due to surface states. Koch and Pincus (1967) have further shown that these states also explain the low field oscillations in surface impedance observed by Koch and Kuo (1967).

## 10.5. Nuclear spin relaxation

When the nuclear spins of a substance are aligned by the application of an external field, they again relax to their equilibrium distribution predominantly by interaction with the conduction electrons. In this interaction, a nucleus flips its spin one way as the electron spin flips the other way so as to conserve the total spin. The electron can do this only if there is available an empty final state of correct energy and spin direction, and the nuclear relaxation rate in a normal metal depends therefore both on the number of conduction electrons (itself proportional to the product of the density of states and the energy derivative of the Fermi function) and on the density of states in the vicinity of the Fermi surface.

To predict the temperature variation of this relaxation process in the superconducting phase one is tempted to use again a simple two-fluid model, according to which the number of 'normal' electrons available decreases rapidly below $T_c$. To this should, therefore, correspond a decrease in the relaxation rate as compared to that in the normal phase. But the energy gap severely modifies the density of states available to the interacting electrons. In the gap there are, by definition, no available states at all, and the missing states are 'piled up' on either side. The presence of the energy derivative of the Fermi function in the relaxation rate expression makes this rate essentially proportional to the square of the density of states evaluated over a range $k_B T$ on either side of the Fermi energy. At temperatures near $T_c$, the gap is still very narrow and $< k_B T$, so that the pile-up of states on either side results in an appreciable increase of the rate over that in the normal phase. At lower temperatures, the gap becomes wider than $k_B T$, and the relaxation rate rapidly diminishes, approaching zero as $T \rightarrow 0$.

The measurements of Hebel and Slichter (1959), of Redfield (1959), and of Masuda and Redfield (1960a) fully confirm this consequence of the energy gap. In particular, a detailed analysis by Hebel (1959) has shown that the empirical results are compatible with the manner of piling up predicted by the microscopic theory. Hebel's results and some empirical values are given in Figure 33, in which the ratio of the relaxation rate in the superconducting phase to that in the normal one is plotted against temperature. The temperature variation of what can be considered as the attenuation of the nuclear alignment is markedly different from the corresponding change in the attenuation of an ultrasonic elastic wave in a superconductor. As this difference is one of the most striking consequences of the BCS theory, its discussion and the general description of ultrasonic attenuation in superconductors is postponed until a later chapter.

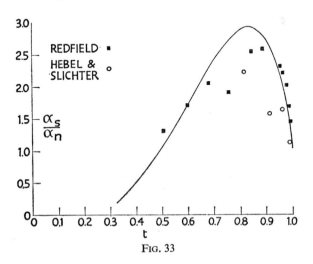

Fig. 33

In his calculations, Hebel avoids singularities on either side of the gap by introducing a parameter $r$ which represents a smearing of the density of states over an energy interval small compared to the width of the gap. It is possible to interpret this in terms of an anisotropy of the gap, since the relaxation process samples the gap over all directions simultaneously. With this interpretation, the data on aluminium of Masuda and Redfield (1960a, 1962) indicate an anisotropy of the order of 1/10 of the gap width, and recent measurement by the same authors (Masuda and Redfield, 1960b; Masuda, 1962b) indicate that this anisotropy decreases in impure aluminium. Anisotropy of magnitude similar to that in aluminium has been found by Masuda (1962a) in cadmium.

## 10.6. The tunnel effect

The most recent and the most direct measurement of the energy gap has been provided by the work of Giaever (1960a), who essentially measured the width of the gap with a voltmeter. He accomplished this by observing the tunneling of electrons between a superconducting film and a normal one across a thin insulating barrier. Quantum-mechanically, an electron on one side of such a barrier has a finite probability of tunneling through it if there is an allowed state of equal

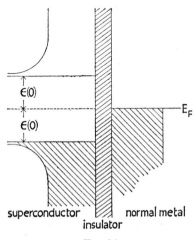

Fig. 34

or smaller energy available for it on the other side. Figure 34 shows the density of states function in energy space for a sandwich consist-ing, from left to right, of a superconductor, an insulator, and a normal metal, all at 0°K. In the last of these, electrons fill all available states up to the Fermi level $E_F$; in the superconductor, there is a gap of half-width $\epsilon(0)$, and states up to $E_F - \epsilon(0)$ are filled. With such conditions there can be no tunneling either way, as on neither side of the barrier are there any available states.

A potential difference applied between the two metals will shift the energy levels of one with respect to the other. It is evident from Figure 34 that tunneling will abruptly become possible when the

9

applied voltage equals $\epsilon(0)$. The subsequent variation of tunneling current with applied voltage of course depends on the details of the density of states curve of the superconductor on either side of the gap. At first, there is a very rapid rise of current with voltage due to the large density of piled-up stages; for voltages much exceeding $\epsilon(0)$, the tunneling samples the density of states well beyond the gap, and the variation of $I$ vs. $V$ approaches the purely ohmic character of a junction of two normal metals. This is summarized in Figure 35, which gives with the solid line the current-voltage characteristic of the superconducting-normal junction at $0°K$. The dotted line indicates the behaviour at $0 < T < T_c$, the modification being due to the fact that at finite temperatures on both sides of the junction some electrons

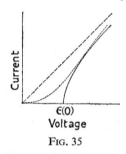

Fig. 35

are excited across the gap or the Fermi level, respectively. The dashed line shows the behaviour at $T > T_c$, i.e. for a junction of normal metals.

Nicol *et al.* (1960) and Giaever (1960b) have extended such experiments to cases where both metals of the junction are superconductors, but with very different critical temperatures, such as Al ($T_c = 1·2°K$) and Pb ($T_c = 7·2°K$). The gaps of the two will be correspondingly different, and for such a junction the density of states function at $0°K$ is shown in Figure 36. A tunneling current will begin to flow when the potential difference between the two metals is $\epsilon(0)_{Pb} + \epsilon(0)_{Al}$. In this case, however, the modification due to finite temperature is more significant than with an $s$–$n$ junction. Imagining the density of states curve of Figure 36 with a few excited electrons beyond both gaps, and a few available states remaining below both, one recognizes that now the current $I$ at first increases with increasing potential $V$, then decreases for $\epsilon(0)_{Pb} - \epsilon(0)_{Al} < V < \epsilon(0)_{Pb} + \epsilon(0)_{Al}$, and then increases again. Figure 37 shows the current-voltage characteristics in this case; the limits of the negative resistance region are very sharp. Thus the current-potential characteristics yield the energy gap values at a given temperature for both metals.

The energy gap values obtained by this method for several superconductors are listed in Table III, and can probably be considered as the most reliable of all experimental determinations. Measurements

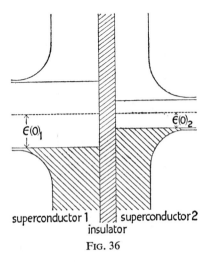

superconductor 1     superconductor 2

insulator

FIG. 36

as a function of temperature closely support the thermal variation of the energy gap predicted by the BSC theory. The films used are thin compared to the penetration depths, and because of their size their critical fields are very high. This and its use in investigating the variation of the energy gap with magnetic field was discussed in Chapter VII. Recent tunneling studies have verified other aspects of the energy gap, in particular its relationship to the phonon spectrum of the superconducting lattice. This will be summarized in Chapter XI.

Simultaneous tunneling of two electrons has been observed by Taylor and Burstein (1962) and Adkins (1963, 1964), in agreement with the calculations of Schrieffer and Wilkins (1962). This is not to be confused with the tunneling of Cooper pairs, as predicted by Josephson (1962), which will be discussed in Section 11.11. The results of Taylor and Burstein also indicate the possibility of tunneling assisted by the simultaneous absorption of a phonon. Theoretical aspects of this have been discussed by Kleinman (1963), Kleinman *et al.* (1964), Fibich (1964). Lax and Vernon (1965) as well as Abeles and Goldstein (1965) have reported measurements on such phonon-assisted tunneling, which is analogous to the photon-assisted tunneling observed by

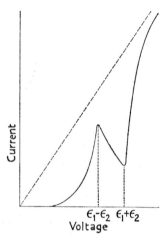

FIG. 37

Dayem and Martin (1962). Calculations by Tien and Gordon (1963) do not seem to be in good agreement with experiments (Goldstein *et al.*, 1966; Cook and Everett, 1967).

Tunneling has been used extensively by the Orsay group, e.g. in studies of 'gapless' superconductivity (Guyon, 1966) and of the magnetic field effect on the gap (Guyon *et al.*, 1965; Orsay group, 1966). This latter work has recently been extended by Levine (1967) and by Millstein and Tinkham (1967).

## 10.7. Far infrared transmission through thin films

In a series of experiments, Tinkham, Glover, and Ginsberg (Glover and Tinkham, 1957; Ginsberg and Tinkham, 1960) have measured the transmission through thin superconducting films of electromagnetic radiation in the far-infrared range of wavelengths between 0·1 and 6 mm. Their results lend themselves to an ingenious analysis leading to a number of very fundamental conclusions about the interrelation of the energy gap, the response to high frequency radiation, and the existence of perfect conductivity and of the Meissner effect in the limit of zero frequency (see Tinkham [10], pp. 168–176).

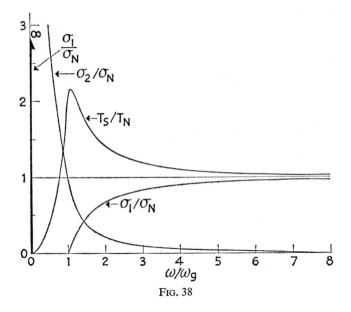

FIG. 38

In Figure 38, the curve labelled $T_s/T_n$ is one which can be drawn through the empirical values of the ratio of the transmissivity in the superconducting phase, $T_s$, to the normal value, $T_n$, all suitably normalized for film resistance and substrate refraction, and plotted against frequency. The transmissivity of a substance is related to its conductivity. One can approximate the conductivity of the film in the normal state by a real number, $\sigma_n$, which to a good approximation is independent of frequency in the range under investigation. The super-conducting conductivity can be written as the complex quantity

$$\sigma_s = \sigma_1(\omega) - i\sigma_2(\omega). \tag{X.2}$$

It then follows from general electromagnetic theory that

$$\frac{T_s}{T_n} = \left\{ \left[ T_n^{1/2} + (1 - T_n^{1/2}) \frac{\sigma_1}{\sigma_n} \right]^2 + \left[ (1 - T_n)^{1/2} \frac{\sigma_2}{\sigma_n} \right]^2 \right\}^{-1}. \tag{X.3}$$

Microwave work on bulk superconductors, such as the measurements of Biondi and Garfunkel (1959), have shown that at $T \ll T_c$ and $\hbar\omega < k_B T_c$, the surface resistance vanishes. It follows from this that the real, lossy part of the conductivity must also vanish in this range, or $\sigma_1 \approx 0$, so that the low frequency measurements of $T_s/T_n$ can be used to evaluate the corresponding values of $\sigma_2/\sigma_n$. For a number of samples of tin and lead with widely varying normal conductivity, all the data of Glover and Tinkham fit a universal curve represented by

$$\sigma_2/\sigma_n = (1/a)(k_B T_c/\hbar\omega), \quad a = 0.27. \tag{X.4}$$

As $\sigma_n$ is independent of frequency, X.4 implies that

$$\sigma_2 \propto 1/\omega. \tag{X.5}$$

This is just the frequency dependence which follows from the London equation

$$\operatorname{curl} \mathbf{J} + \frac{c}{4\pi\lambda^2} \mathbf{H} = 0, \tag{X.6}$$

since this with Maxwell's equation

$$\operatorname{curl} \mathbf{E} = \frac{1}{c} \dot{\mathbf{H}}$$

leads to

$$\sigma_2 = \frac{c^2}{4\pi\lambda^2} \frac{1}{\omega}. \tag{X.7}$$

An imaginary conductivity which is inversely proportional to the frequency thus corresponds to the consequences of X.6: the Meissner effect and a finite penetration depth $\lambda$. However, the magnitude of $\lambda$

calculated from the experimental transmission results with the aid of X.7 exceeds the London value $\lambda_L = mc^2/4\pi^2ne^2$ by at least a factor of ten. Furthermore, there is nothing in the London theory to explain why $\sigma_2/\sigma_n$ for different superconductors should satisfy a universal equation like X.4. On the other hand the Pippard treatment predicts for these films, in which $\xi \approx l \ll \lambda$, that (see equation IV.18a):

$$\lambda^2 = (\xi_0/l)\lambda_L^2,$$

where (equation IV.9):

$$\xi_0 = a\hbar v_0/k_B T_c.$$

Hence

$$\frac{\sigma_2}{\sigma_n} = \frac{ne^2}{m} \times \frac{l}{\xi_0} \times \frac{1}{\omega} \times \frac{1}{\sigma_n}.$$

For a normal metal

$$\frac{\sigma_n}{l} = \frac{ne^2}{mv_0},$$

and hence the Pippard theory leads to

$$\frac{\sigma_2}{\sigma_n} = \frac{1}{a}\frac{k_B T_c}{\hbar\omega}$$

for all superconductors. This is strikingly verified by the results of Glover and Tinkham.

The real and imaginary parts of any linear response function, such as the electrical conductivity, are related by a pair of integral transforms known as the Kramers-Kronig (K-K) relations. In terms of the conductivity these take the form:

$$\sigma_1(\omega) = \frac{1}{\pi}\int_{-\infty}^{+\infty}\frac{\omega_1\,\sigma_2(\omega_1)\,d\omega_1}{\omega_1^2 - \omega^2}; \quad \sigma_2(\omega) = -\frac{\omega}{\pi}\int_{-\infty}^{+\infty}\frac{\sigma_1(\omega_1)\,d\omega_1}{\omega_1^2 - \omega^2}. \quad \text{(X.8)}$$

Substituting X.7 into the first of these two relations shows that the imaginary conductivity $\sigma_2$ must be accompanied by a real conductivity which takes the form of a delta-function at the origin:

$$\sigma_1(\omega) = (c^2/8\lambda^2)\,\delta(\omega - 0). \tag{X.9}$$

Similarly, in terms of the empirical value X.4 for $\sigma_2/\sigma_n$ one would have

$$\frac{\sigma_1}{\sigma_n} = \frac{\pi}{2}\frac{1}{a}\frac{k_B T_c}{\hbar}\delta(\omega - 0). \tag{X.10}$$

Such an infinite real conductivity at zero frequency of course does not introduce losses.

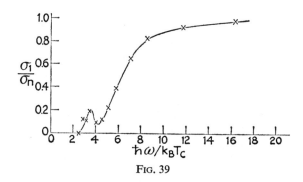

FIG. 39

Turning now to the high frequency far-infrared transmissivity data the peak and subsequent decrease of $T_s/T_n$ indicates that at a frequency roughly corresponding to the peak, a real, lossy component $\sigma_1$ of the superconducting conductivity must appear. In the absence of such a component $T_s/T_n$ would continue to rise. The appearance of a real component of conductivity at or near some critical frequency is, of course, highly suggestive of an energy gap. Taken by themselves, the data of Tinkham *et al.* do not determine the gap quite unambiguously (see Forrester, 1958). However, accepting the existence of a gap from other experiments allows a fully consistent interpretation of the transmission results from which the magnitude of the gap as well as other interesting quantities can be derived.

The calculations of Miller (1960) of the variation of $\sigma_1/\sigma_n$ are shown in Figure 38, the ordinate being scaled in units of

$$\omega_g \equiv \frac{2\epsilon(0)}{\hbar},$$

where $2\epsilon(0)$ is the width of the gap at $0°K$. An energy gap implies that, as for a normal metal, the imaginary part of the conductivity vanishes for frequencies beyond the gap. Using X.3 one can then calculate $\sigma_1/\sigma_n$ from the measured values of $T_s/T_n$ to a first approximation, and then apply an iterative procedure using the K-K relations as well as the sum rule about to be mentioned to obtain final values of $\sigma_1/\sigma_n$. Figure 39 gives the result thus obtained by Ginsberg and Tinkham for lead, showing the precursor peak also found for mercury. One ignores this in deriving energy gap values from the limit $\sigma_1/\sigma_n \to 0$. The resulting gap widths are listed in Table III.

### 10.8. The Ferrell-Glover sum rule

The intimate connection between the experimentally verified decrease of $\sigma_1/\sigma_n$ near $\omega_g$, corresponding to the existence of a gap, and the low frequency London-type imaginary conductivity $\sigma_2^L \propto 1/\omega$, corresponding to infinite conductivity and the Meissner effect at zero frequency, was first pointed out by Ferrell and Glover (1958) and further elaborated by Tinkham and Ferrell (1959). The first of these papers pointed out that at extremely high frequencies, such that $\hbar\omega$ far exceeds any of the binding energies of an electron in the metal, the real part of the conductivity vanishes. The appropriate K-K relation for the imaginary conductivity then becomes, since $\sigma_1$ is an even function,

$$\sigma_2(\omega) \approx \frac{2}{\pi\omega} \int\limits_0^\infty \sigma_1(\omega_1)\,d\omega_1. \tag{X.11}$$

At these very high frequencies all electrons are free in both the normal and the superconducting phases, and one would thus expect $\sigma_2(\omega)$ and, therefore, the integral in X.11 to have the same value in both phases. In other words, there exists the sum rule that this integral remains unchanged under the superconducting transition. From this it follows that any area $A$ removed from under the $\sigma_1(\omega)$ curve by the energy gap must reappear somewhere else, and it can do so only at the origin in the form of a delta function of strength $A$. This being the case, one can then again apply the K-K relations to show that associated with such a delta function

$$\sigma_1(\omega) = A\delta(\omega - 0) \tag{X.12}$$

must be a contribution to the imaginary conductivity of magnitude

$$\sigma_2^A(\omega) = 2A/\pi\omega. \tag{X.13}$$

The argument has now come through a full circle. An energy gap corresponds to a disappearance of $\sigma_1(\omega)$ in the superconducting phase over some frequency range in which this conductivity is finite in the normal metal. This, according to the Ferrel-Glover sum rule, must lead to the appearance of a delta function X.12. In turn this leads to a London-type imaginary conductivity $\sigma_2 \propto 1/\omega$, which was seen to correspond to the Meissner effect and infinite conductivity. One sees further that in terms of the parameter $a$ of X.4, one can write

$$A/\sigma_n = (\pi/2)(k_B T_c/\hbar)(1/a). \tag{X.14}$$

Determining $A/\sigma_n$ from their transmission data and using this relation, Ginsberg and Tinkham obtain values for $a$ of $0.23$ for lead, $0.26$ for tin, and $0.19$ for thallium. These, as well as Glover and Tinkham's value of $0.27$ for both tin and lead, are in remarkable agreement both with the Faber-Pippard data ($0.15$ for tin and indium) and with the BCS prediction for all metals ($0.18$). The agreement is particularly convincing if one considers the simplifications of the theory on the one hand, and the wide variety and considerable difficulty of the experiments on the other.

From X.13 and X.7 it is evident that

$$\lambda^2 = c^2/8A. \tag{X.15}$$

Thus the Ferrell-Glover sum-rule leads to an inverse proportionality between the square of the penetration depth and the energy gap. Such a relation is implicit in the Pippard model and the Bardeen theory, and appears explicitly in the Ginzburg-Landau treatment as extended by Gor'kov.

# CHAPTER XI

# Microscopic Theory of Superconductivity

## 11.1. Introduction

In reviewing the contents of the preceding chapters, which give an empirical description of superconductivity, perhaps the most striking feature to be noticed is how much quantitative information can be given about superconductivity in general without speaking about the specific properties of any one of the many superconducting elements. The astonishing degree of similarity in the superconducting behaviour of metals with widely varying crystallographic and atomic properties indicates that the explanation for superconductivity should be inherent in a general, idealized model of a metal which ignores the complicated features characterizing any individual metallic element. It should, therefore, be possible to find in the simple model of the ideal metal the possibility of an interaction mechanism leading to the superconducting state, and to derive from this at least qualitatively the properties of an ideal superconductor.

One would judge from this that an explanation for superconductivity should be fairly easy, until he realizes the extreme smallness of the energy involved. A superconductor can be made normal by the application of a magnetic field $H_c$ which at absolute zero is of the order of a few hundred gauss. The energy difference between the superconducting and the normal phase at absolute zero, which is given by $H_0^2/8\pi$, thus is of the order of $10^{-8}$ e.v. per atom. How very small this is can best be judged by remembering that the Fermi energy of the conduction electrons in a normal metal is of the order of 10–20 e.v. The simple model of Bloch and Sommerfeld gives a reasonably accurate description of the basic characteristics of a metal although it completely ignores, among other things, the correlation energy of the conduction electrons due to their Coulomb interaction. This energy is of the order of 1 e.v.!

As a further difficulty in arriving at a microscopic theory of superconductivity one must add the extreme sharpness of the phase transition under suitable conditions. The absence of statistical fluctuations shows that the superconducting state is a highly correlated one involving a very large number of electrons. Thus it is necessary to find inherent in the basic properties common to all metals an interaction correlating a large number of electrons in such a way that the energy of the system relative to the normal metal is lowered by a very small amount. The discovery of the isotope effect in a number of superconducting elements clearly indicated that in these the interaction in question must be one between the electrons and the vibrating crystal lattice, and indeed Fröhlich (1950) had suggested such a mechanism independently of the simultaneous experimental results.

## 11.2. The electron–phonon interaction

Fröhlich and, a little later, Bardeen (1950) pointed out that an electron moving through a crystal lattice has a self energy by being 'clothed' with virtual phonons. What this means is that an electron moving through the lattice distorts the lattice, and the lattice in turn acts on the electron by virtue of the electrostatic forces between them. The oscillatory distortion of the lattice is quantized in terms of phonons, and so one can think of the interaction between lattice and electron as the constant emission and reabsorption of phonons by the latter. These are called 'virtual' phonons because as a consequence of the uncertainty principle their very short lifetime renders it unnecessary to conserve energy in the process. Thus one can think of the electron moving through the lattice as being accompanied or 'clothed', even at $0°K$, by a cloud of virtual phonons. This contributes to the electron an amount of self-energy which, as was pointed out by Fröhlich and by Bardeen, is proportional to the square of the average phonon energy. In turn this is inversely proportional to the lattice mass, so that a condensation energy equal to this self-energy would have the correct mass dependence indicated by the isotope effect. Unfortunately, however, the size turns out to be three to four orders of magnitude too large.

It was only seven years later that Bardeen, Cooper, and Schrieffer (BCS, 1957) succeeded in showing that the basic interaction responsible for superconductivity appears to be that of a *pair* of electrons by means of an interchange of virtual phonons. In the simple terms used above this means that the lattice is distorted by a moving electron, this distortion giving rise to a phonon. A second electron some distance away is in turn affected when it is reached by the propagating fluctuation in the lattice charge distribution. In other words, as shown in Figure 40, an electron of wave vector **k** emits a virtual phonon **q** which is absorbed by an electron **k′**. This scatters **k** into **k − q** and **k′** into **k′ + q**. The process being a virtual one, energy need not be conserved, and in fact the nature of the resulting electron–electron interaction depends on the relative magnitudes of the electronic energy change and the phonon energy $\hbar\omega_q$. If this latter exceeds the

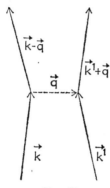

FIG. 40

former, the interaction is attractive—the charge fluctuation of the lattice is then such as to surround one of the electrons by a positive screening charge greater than the electronic one, so that the second electron sees and is attracted by a net positive charge.

The fundamental postulate of the BCS theory is that superconductivity occurs when such an attractive interaction between two electrons by means of phonon exchange dominates the usual repulsive screened Coulomb interaction.

## 11.3. The Cooper pairs

Shortly before the formulation of the BCS theory, Cooper (1956) had been able to show that if there is a net attraction, however weak, between a pair of electrons just above the Fermi surface, these electrons can form a bound state. The electrons for which this can occur as a result of the phonon interaction lie in a thin shell of width $\simeq \hbar\omega_q$, where $\hbar\omega_q$ is of the order of the average phonon energy of the metal. If one looks at the matrix elements for all possible interactions which take a pair of electrons from any two **k** values in this shell to any two others, he finds that because of the Fermi statistics of the electron these matrix elements alternate in sign and, being all of roughly equal magnitude, give a negligible total interaction energy, that is, a vanishingly small total lowering of the energy relative to the normal situation of unpaired electrons. One can, however, restrict oneself to matrix elements of a single sign by associating all possible **k** values in pairs, $\mathbf{k}_1$ and $\mathbf{k}_2$, and requiring that either both or neither member of a pair be occupied. As the lowest energy is obtained by having the largest number of possible transitions, each represented by a matrix element all of the same sign, one wants to choose these pairs in such a way that from any one set of values $(\mathbf{k}_1, \mathbf{k}_2)$, transitions are possible into all other pairs $(\mathbf{k}_1', \mathbf{k}_2')$. As momentum must be conserved, this means that one must require that

$$\mathbf{k}_1 + \mathbf{k}_2 = \mathbf{k}_1' + \mathbf{k}_2' = \mathbf{K} \tag{XI.1}$$

that is, that all bound pairs should have the same total momentum **K**. (See, for example, Cooper, 1960.)

To find the possible value of $\mathbf{k}_1$ and $\mathbf{k}_2$ which satisfy XI.1 and at the same time lie in a narrow shell straddling the Fermi surface $\mathbf{k}_F$ one can construct the diagram shown in Figure 41, drawing concentric circles of radii $k_F - \delta$ and $k_F + \delta$ from two points separated by **K**. It is clear that all possible values of $\mathbf{k}_1$ and $\mathbf{k}_2$ satisfying XI.1 are restricted to the two shaded regions. This shows that the volume of phase space available for what has become known as Cooper pairs has a very sharp maximum for $\mathbf{K} = 0$. Thus the largest number of possible transitions yielding the most appreciable lowering of energy is obtained by pairing all possible states such that their total momentum vanishes. It is also possible to show that exchange terms tend to reduce the interaction energy for pairs of parallel spin, so that it is energetically most favourable to restrict the pairs to those of opposite spin. One can, therefore, summarize the basic hypothesis of the BCS theory as follows: *At 0°K the superconducting ground state is a highly correlated one in which in momentum space the normal electron states in a thin shell near the Fermi surface are to the fullest extent possible occupied by pairs of opposite spin and momentum.* The most direct verification of the existence of these pairs arises from the flux quantization measurements mentioned in Chapter III.

The energy of this state is lower than that of the normal metal by a finite amount which is the condensation energy of the superconducting state and which at 0°K must equal $H_0^2/8\pi$ per unit volume. Furthermore, this state has the all-important property that it takes a finite quantity of energy to excite even a single 'normal', unpaired electron. For not only does this require the very small amount of energy needed

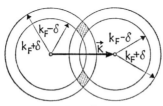

Fig. 41

to break up a bound pair, but more importantly the occupation of a single **k** state by an unpaired electron removes from the system a large number of pairs which could have interacted so as to occupy **k** and −**k**. Hence the total energy difference between having all paired electrons and having a single excited electron is finite and equal to a large multiple of the single pair correlation energy. In terms of the single electron spectrum, therefore, the BCS theory correctly yields an energy gap. It has already been shown that such an energy gap not only leads to the observed variation of the specific heat, the thermal conductivity, and the absorption of high frequency electromagnetic radiation, but also that it is correlated with the existence of perfect diamagnetism and perfect conductivity in the low frequency limit.

## 11.4. The ground state energy

The recognition of the basic electron interaction mechanism responsible for superconductivity does not remove the major difficulty mentioned earlier, namely that the correlation energy in question is so very much smaller than almost any other contribution to the total electronic energy. BCS therefore take the bold step of assuming that all interactions except the crucial one are the same for the superconducting as for the normal ground state at 0°K. Taking as the zero of energy the normal ground state energy and including in this all normal state correlations and even the self energy of the electrons due to virtual phonon emission and reabsorption, BCS proceed to calculate the superconducting ground state energy as being due uniquely to the correlation between Cooper pairs of electrons of opposite spin and momentum by phonon and screened Coulomb interaction.

The interaction leading to the transition of a pair of electrons from the state $(\mathbf{k}\uparrow, -\mathbf{k}\downarrow)$ to $(\mathbf{k}'\uparrow, -\mathbf{k}'\downarrow)$ is characterized by a matrix element,

$$-V_{kk'} = 2(-\mathbf{k}'\downarrow, \mathbf{k}'\uparrow |H_{int}| -\mathbf{k}\downarrow, \mathbf{k}\uparrow), \qquad \text{(XI.2)}$$

where $H_{int}$ is the truncated Hamiltonian from which all terms common to the normal and superconducting phases have been removed. $V_{kk'}$ is the difference between one term describing the interaction between the two electrons by means of a phonon, and a second one giving their screened Coulomb interaction. The basic similarity of the superconducting characteristics of widely different metals implies that the responsible interaction cannot crucially depend on details characteristic of individual substances. BCS therefore make the further simplifying assumption that $V_{kk'}$ is isotropic and constant for all electrons in a narrow shell, straddling the Fermi surface, of thickness (in units of energy) less than the average energy of the lattice, and that $V_{kk'}$ vanishes elsewhere. Measuring electron energy from the Fermi surface, and calling $\epsilon_k$ the energy of an electron in state $k$, one can state this formally by the equations:

$$\left.\begin{array}{l} V_{kk'} = V \quad \text{for } |\epsilon_k|, |\epsilon_{k'}| \leqslant \hbar\omega_q, \\[2mm] V_{kk'} = 0 \text{ elsewhere.} \end{array}\right\} \text{(XI.4)}$$

and

The basic BCS criterion for superconductivity is equivalent to the condition

$$V < 0.$$

It is well to note clearly at this point that this simplification of the interaction parameter $V$ necessarily leads to what can be called a law of corresponding states for all superconductors, that is, virtually identical predictions for the magnitudes of all characteristic quantities in terms of reduced co-ordinates. Any empirical deviation from such complete similarity is, therefore, no invalidation of the basic premise of the BCS theory, but merely an indication of the oversimplification inherent in XI.4. (See footnote, page 143.)

Let $h_k$ be the probability that states $\mathbf{k}$ and $-\mathbf{k}$ are occupied by a pair of electrons, and $(1 - h_k)$ the corresponding probability that the states are empty. $W(0)$, the ground state energy of the superconducting state at $0°$K as compared to the energy of the normal metal, is then given by

$$W(0) = \sum_k 2\epsilon_k h_k - \sum_{kk'} V_{kk'}\{h_k(1-h_{k'})h_{k'}(1-h_k)\}^{1/2}. \quad (XI.5)$$

The summation is over all those $\mathbf{k}$-values for which $V_{kk'} \neq 0$, so that using XI.4 one can simplify to

$$W(0) = \sum_k 2\epsilon_k h_k - V \sum_{kk'} \{h_k(1-h_{k'})h_{k'}(1-h_k)\}^{1/2}. \quad (XI.5')$$

The first term gives the difference of kinetic energy between the superconducting and normal phases at $0°$K. The factor 2 arises because for every electron in state $\mathbf{k}$ of energy $\epsilon_k$ there is with an isotropic Fermi surface another electron of the same energy in $-\mathbf{k}$. This first term can be either positive or negative, and gives the correlation energy for all possible transitions from a pair state $(\mathbf{k}, -\mathbf{k})$ to another $(\mathbf{k}', -\mathbf{k}')$. For such a transition to be possible, $\mathbf{k}$ must initially be occupied and $\mathbf{k}'$ empty. The simultaneous probability of this is given by $h_k(1 - h_{k'})$. The final state must have $\mathbf{k}$ empty and $\mathbf{k}'$ occupied, and this has probability $h_{k'}(1-h_k)$. The square root of the product of these probabilities multiplied by the matrix element for the transition and summed over all possible values of $\mathbf{k}$ and $\mathbf{k}'$ gives the total correlation energy.

$W(0)$ must of course be negative for the superconducting phase to

10

exist, and to see whether this is possible XI.5' can be minimized with respect to $h_k$. This leads to

$$\frac{[h_k(1-h_k)]^{1/2}}{1-2h_k} = V \frac{\sum\limits_{k'} [h_{k'}(1-h_{k'})]^{1/2}}{2\epsilon_k}. \tag{XI.6}$$

By defining

$$\epsilon(0) \equiv V \sum_{k'} [h_{k'}(1-h_{k'})]^{1/2}, \tag{XI.7}$$

equation XI.6 simplifies to

$$h_k = \frac{1}{2}\left(1 - \frac{\epsilon_k}{E_k}\right), \tag{XI.8}$$

where

$$E_k \equiv [\epsilon_k^2 + \epsilon^2(0)]^{1/2}. \tag{XI.9}$$

Substituting XI.8 back into XI.7 one obtains a non-linear relation for $\epsilon(0)$:

$$\epsilon(0) = \frac{V}{2} \sum_k \frac{\epsilon(0)}{[\epsilon_k^2 + \epsilon^2(0)]^{1/2}}. \tag{XI.10}$$

This can be treated most readily by changing the summation to an integration and transforming the variable of integration from **k** to $\epsilon$. Assuming symmetry of states on either side of the Fermi surface ($\epsilon = 0$), and introducing the density of single electron states of one spin in the normal state at $\epsilon = 0$: $N(0)$, XI.10 becomes

$$\frac{1}{N(0)\,V} = \int_0^{\hbar\omega_q} \frac{d\epsilon}{[\epsilon^2 + \epsilon^2(0)]^{1/2}}. \tag{XI.11}$$

The limit of integration is the phonon energy above which, according to XI.4, $V = 0$.

The solution of XI.11 is

$$\epsilon(0) = \hbar\omega_q/\sinh[1/N(0)\,V]. \tag{XI.12}$$

Putting this back into XI.9 and XI.7 and finally into XI.5', one finds that the ground state energy of the superconducting state is given by

$$W(0) = -\frac{2N(0)\,(\hbar\omega_q)^2}{\exp[2/N(0)\,V] - 1}. \tag{XI.13}$$

The numerator of this quantity follows from dimensional reasoning from any theory which postulates an interaction between electrons and phonons and allows this interaction to be cut off at some average phonon energy $\hbar\omega_q \approx k_B\Theta$, beyond which the interaction becomes repulsive. A term like this had been contained in the earlier attempts of Fröhlich and of Bardeen, and, as mentioned before, is much too large. The success of the BCS theory lies in the appearance of the exponential denominator which reduces $W(0)$ by many orders of magnitude. Although a precise calculation of the average interaction parameter $V$ for a specific metal continues to be among the most important questions still to be solved, various estimates (Pines, 1958; Morel, 1959; Morel and Anderson, 1962) indicate that the values of $N(0)V \approx 0.3$, derived from a knowledge of $H_0$, are reasonable. Thus the denominator has a value of about $e^7$.

The isotope effect follows from the numerator of XI.13, as it would from any theory involving electron–phonon interaction with a cut-off frequency related to the Debye $\Theta$ and hence to the isotopic mass. Equation XI.13 shows that

$$\frac{H_0^2}{8\pi} = W(0) \propto (\hbar\omega_q)^2 \approx (k_B\Theta)^2 \propto M_{\text{ion}}^{-1}. \qquad \text{(XI.14)}$$

For a group of isotopes, one finds $H_0 \propto T_c$, so that

$$T_c \propto M_{\text{ion}}^{-1/2}. \qquad \text{(XI.15)}$$

Any appreciable deviation of the isotope effect exponent from the value 0.5 could indicate that the simplifying BCS assumption of a cut-off for both Coulomb and phonon interaction at $\hbar\omega_q$ has to be modified (Tolmachev, 1958; Swihart, 1959, 1962). Bardeen (1959) has pointed out that the cut-off may be determined by the lifetime of the 'normal' electrons which can be excited across the gap. These electrons are not the bare, non-interacting electrons of the simple Bloch-Sommerfeld model. Instead they are so-called quasi-particles 'clothed' by their interactions with each other and with the lattice (see [7], pp. 184–95). As a result the wave functions describing them are not eigenfunctions of the system, so that the particles have a finite lifetime. The effect of this on the pair interaction has been further discussed by Eliashberg (1961), Bardeen [9], Schrieffer (1961), Betbeder-Matibet and Nozières (1961), and Bardasis and Schrieffer (1961). The damping of the quasi-particles is found to be very small even up to energies well beyond the Fermi energy. This is in contradiction to the BCS assumption embodied in XI.4, as the justification of the cut-off of the Coulomb interaction at $\hbar\omega_q$ is essentially that quasi-particles of larger energy are so strongly damped as not to be available for pair formation. It is thus necessary to modify the BCS cut-off by taking into account the existence of the repulsive interaction for $\epsilon_k > \hbar\omega_q$. This does not appreciably affect the gap at the Fermi surface ($\epsilon_k = 0$), but will result in its variation with $\epsilon_k$, as will be further discussed in Section 11.7. The theory of strong coupling superconductors described there is based on the Eliashberg interaction potential, which has now been generally accepted as the correct one.

With a compound tunneling arrangement in which electrons are injected into a layer of superconducting lead and then have the possibility of tunneling through a second junction into normal metal, Ginsberg (1962) was recently able to place an upper limit on the lifetime of the quasi-particles in a superconductor. According to his preliminary result this upper bound is $2 \cdot 2 \times 10^{-7}$ sec, which is only about five times as large as the average time calculated by Schrieffer and Ginsberg (1962) for quasi-particle recombination into pairs by means of phonon emission. This has also been calculated by Rothwarf and Cohen (1963).

Swihart (1962) as well as Morel and Anderson (1962) have studied the isotope effect for different forms of the energy dependence of the electron-electron interaction. They find that the exponent of the isotopic mass in equation XI.15 is *less* than the ideal value of one half by amounts of 10–30 per cent which increase with decreasing $T_c/\Theta$. However, the isotope effects in ruthenium (Geballe *et al.*, 1961, Finnemore and Mapother, 1962), osmium (Hein and Gibson, 1964) and perhaps also in molybdenum (Matthias *et al.*, 1963) appear to be too small to be explained by these calculations.

This raises questions about the origin of the attractive interaction responsible for the formation of Cooper pairs in these as well as perhaps in other metals. Matthias (see for example, 1960) has repeatedly suggested that in all transition metals there exists an attractive magnetic interaction responsible for superconductivity. However, both Kondo (1963) and Garland (1963a) have tried to explain the apparently anomalous superconducting behaviour of the transition metals as a consequence of the overlap at the Fermi energy of the *s* and *d* bands of the electronic spectrum, and not because of a magnetic interaction. Kondo assumes a larger interband interaction; Garland, on the other hand, believes that the electrons of high effective mass in the *d*-band tend not to follow the motion of the *s*-electrons. This results in 'anti-shielding' the interactions between *s*-electrons, leading to an attractive screened Coulomb interaction between them being added to the usual attractive interaction by exchange of virtual phonons.

Garland (1963b) calculated the magnitude of the isotope effect for all superconducting elements and obtains results which agree closely with all available experimental results, including in particular the reduced effect in transition metals. This also results, at least qualitatively, from Kondo's calculations. Garland was also able to explain the anomalous pressure effect in transition metals (Bucher and Olsen, 1964).

### 11.5. The energy gap at $0°K$

From XI.5′ one can see that the contribution of a single pair state $(\mathbf{k}, -\mathbf{k})$ to this total condensation energy is

$$W_k = 2\epsilon_k h_k - 2V \sum_{k'} \{(1-h_{k'})h_{k'}\}^{1/2}. \qquad \text{(XI.16)}$$

The first term represents the kinetic energy of both electrons in the pair state $\mathbf{k}$, and the second term the total interaction energy due to all possible transitions into or out of the state.

At $0°K$ the lowest excited state of the superconductor must correspond to breaking up a single pair by transferring an electron from a state $\mathbf{k}$ to another, leaving an unpaired electron in $-\mathbf{k}$. The condensation energy is then reduced by $W_k$. The first term of this can be made arbitrarily small, and is analogous to the excitation energy in a normal metal, for which there is a quasi-continuous energy spectrum above the ground state. The second term of $W_k$, however, is finite for all values of $\mathbf{k}$, which is why in the superconducting phase the lowest excited state is separated from the ground state by an energy gap.

Comparing XI.16 with XI.7 one sees that this energy gap has the value $2\epsilon(0)$, which according to XI.12 equals

$$2\epsilon(0) = 2\hbar\omega_q/\sinh[1/N(0)\,V].\qquad\text{(XI.17)}$$

As $1/N(0)\,V \approx 3-4$, this can be approximated by

$$2\epsilon(0) = 4\hbar\omega_q\exp[-1/N(0)\,V].\qquad\text{(XI.18)}$$

### 11.6. The superconductor at finite temperatures

As the temperature of the superconductor is raised above $0°\text{K}$, an increasing number of electrons find themselves thermally excited into single quasi-particle states. These excitations behave like those of a normal metal; they are readily scattered and can gain or lose further energy in arbitrarily small quantities. In what follows they are simply called normal electrons. At the same time there continues to exist the configuration of all electrons still correlated into Cooper pairs, and displaying superconducting properties, being very difficult to scatter or to excite. One is thus led again to a two-fluid point of view.

As at $0°\text{K}$, one can write down an analytic expression for the ground state energy $W(T)$ containing a kinetic energy term and an interaction term. In both, the presence of the normal electrons must be accounted for, which is done by introducing a suitable probability factor $f_k$.

Letting $f_k$ = probability of occupation of $\mathbf{k}$ or of $-\mathbf{k}$ by a single normal electron, then

$1-2f_k$ = probability that neither $\mathbf{k}$ nor $-\mathbf{k}$ is occupied by a normal electron.

This leads to a kinetic energy term

$$[W(T)]_{\text{K.E.}} = 2\sum_k |\epsilon_k|\,[f_k+(1-2f_k)\,h_k],\qquad\text{(XI.19)}$$

where the summation is over the same range as at $0°\text{K}$, and $h_k$ retains the same definition, though no longer the same value. The second term in the brackets clearly gives the probability that the pair state $(\mathbf{k}, -\mathbf{k})$ not occupied by normal electrons but by a correlated pair. The correlation energy at a finite temperature is

$$[W(T)]_{\text{corr}} = -V\sum_{kk'}\{h_k(1-h_{k'})\,h_{k'}(1-h_k)\}^{1/2}\times$$
$$\times(1-2f_k)(1-2f_{k'}).\qquad\text{(XI.20)}$$

The last two terms ensure that the correlated pair states not be occupied by normal electrons. It is obvious that the presence of these terms decreases the pairing energy.

The thermal properties of the superconductors can now be found quite readily by writing down the free energy of the system and requiring this to be at a minimum. The free energy is

$$G = W(T) - TS = [W(T)]_{\text{K.E.}} + [W(T)]_{\text{corr}} - TS, \quad \text{(XI.21)}$$

where $T$ is the temperature and $S$ the entropy. This last is due entirely to the normal electrons; the electrons which are still paired are in a state of highest possible order and do not contribute at all. Thus the entropy is given by the usual expression for particles obeying Fermi-Dirac statistics:

$$TS = -2k_B T \sum_k \{ f_k \ln f_k + (1 - f_k) \ln (1 - f_k) \}. \quad \text{(XI.22)}$$

Substituting XI.19, XI.20 and XI.22 into XI.21, and minimizing this free energy with respect to $h_k$, one now obtains

$$\frac{[h_k(1 - h_k)]^{1/2}}{1 - 2h_k} = V \frac{\sum_{k'} [h_{k'}(1 - h_{k'})]^{1/2}(1 - 2f_{k'})}{2\epsilon_k}. \quad \text{(XI.23)}$$

This time one defines

$$\epsilon(T) \equiv V \sum_{k'} [h_{k'}(1 - h_{k'})]^{1/2}(1 - 2f_{k'}), \quad \text{(XI.24)}$$

and again obtains

$$h_k = \frac{1}{2} \left[ 1 - \frac{\epsilon_k}{E_k} \right], \quad \text{(XI.25)}$$

where $E_k$ is now defined as $E_k \equiv [\epsilon_k^2 + \epsilon^2(T)]^{1/2}$.

One sees that, as at $0°K$, $2\epsilon(T)$ represents the contribution of a single pair state to the total correlation energy, and that to break up one such pair at any finite temperature removes from the superconducting energy at least this amount. In other words, the superconducting state continues to contain an energy gap $2\epsilon(T)$ separating the lowest energy configuration at any given temperature from that with one less correlated pair.

To evaluate the magnitude of the energy gap one must first find an expression for $f_k$, which one obtains by minimizing the free energy with respect to $f_k$. This yields

$$f_k = [\exp(E_k/k_B T) + 1]^{-1}. \quad \text{(XI.26)}$$

XI.26, XI.18, and XI.24 yield for $\epsilon(T)$ a non-linear relation which, changing as before from a summation over **k** to an integration over $\epsilon$, becomes

$$\frac{1}{N(0)\,V} = \int\limits_{0}^{\hbar\omega_q} \frac{d\epsilon}{[\epsilon^2 + \epsilon^2(T)]^{1/2}} \tanh\left\{ \frac{[\epsilon^2 + \epsilon^2(T)]^{1/2}}{2k_B T} \right\}. \quad \text{(XI.27)}$$

The critical temperature $T_c$ is reached when all pair states are broken up so that $\epsilon(T_c) = 0$. Hence

$$\frac{1}{N(0)\,V} = \int\limits_{0}^{\hbar\omega_q} \frac{d\epsilon}{\epsilon} \tanh \frac{\epsilon}{2k_B T_c}. \quad \text{(XI.28)}$$

As long as $k_B T_c \ll \hbar\omega_q$, the solution of this can be written as

$$k_B T_c = 1\cdot14\,\hbar\omega_q \exp[-1/N(0)\,V]. \quad \text{(XI.29)}$$

The exponential dependence of the transition temperature has been verified by Olsen *et al.* (1964) by means of measurements of its variation with pressure in aluminium.

### 11.7. Experimental verification of predicted thermal properties

Combining equations XI.18 and XI.29 yields for the width of the energy gap at $0°K$

$$2\epsilon(0) = 3\cdot52\,k_B T_c. \quad \text{(XI.30)}$$

This is in remarkable quantitative agreement with empirical values obtained from the wide variety of measurements mentioned in Chapter X. Table III shows that for the most widely different elements the energy gap does not appear to deviate from this idealized value by more than about 20 per cent. The theoretical temperature variation of the gap width is displayed in Figure 42; this has also been well confirmed by a number of experiments.

Mühlschlegel (1959) has tabulated values of the energy gap, the entropy, the critical magnetic field, the penetration depth, and the specific heat, all in reduced coordinates, as functions of the reduced temperature. All these are in close agreement with experimental results.

These agreements clearly vindicate the basic BCS approach, according to which the similarities between superconductors outweigh their differences, so that an approximate law of corresponding states should

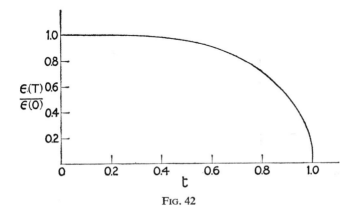

FIG. 42

hold.† This similarity principle had of course emerged from much previous experimental evidence. However, differences between metals and anisotropies in a given metal do exist, and the experimental evidence for gap variations from one metal to another, as well as for gap anisotropies, clearly indicates the need to refine the details of the BCS calculations. For one thing it is of course desirable to take into account

† Deviations from such a law can occur even with the BCS assumption of constant $V$ if in solving equations XI.27 and XI.28 one takes into account higher order terms in $k_B T_c/\hbar\omega_q$ (Mühlschlegel, 1959). The resulting correction factors appearing in equations XI.30, XI.35, and XI.36 are, however, too small to explain the empirical deviations from similarity discussed in this section. Thouless (1960) has shown that in the BCS formulation the energy gap at $0°K$ is only $4\cdot0\, k_B T_c$ even in the non-physical limit

$$k_B T_c/\hbar\omega_q \to \infty.$$

the dependence of the interaction parameter $V$ on $\mathbf{k}$ and $\mathbf{k}'$, so as to be able to calculate directional effects. Even more challenging are the previously mentioned attempts to relax, even in an isotropic model, the assumption XI.4 that $V$ is strictly constant for $\epsilon_k \leqslant \hbar\omega_q$ and is then cut off abruptly. A better knowledge of the variation of $V$ with $\epsilon_k$ in turn would allow the more precise calculation of the corresponding dependence of the energy gap on $\epsilon_k$. The actual form of this variation undoubtedly more nearly resembles the solid line in Figure 43 rather than the dotted line which corresponds to the simple BCS assumption. Usually one is interested in excitation energies of the order of $k_B T_c$ and the BCS assumption is then fully applicable as long as $k_B T_c \ll \hbar\omega_q$,

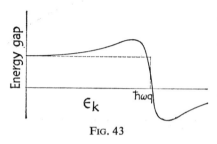

FIG. 43

which is called the weak coupling limit. As $\hbar\omega_q \approx k_B\Theta$, where $\Theta$ is the Debye temperature, this requires that

$$T_c \ll \Theta.$$

For a number of superconducting elements, in particular for Pb and Hg, this condition does not hold. These are called *strong coupling* superconductors.

The first attempts with strong coupling superconductors to use a more realistic energy dependence of the interaction $V$ than that assumed in the BCS theory were undertaken by Swihart (1962, 1963) and by Morel and Anderson (1962). These authors took into account that, as mentioned earlier (see Section 11.4), quasi-particle damping is too small to justify cutting off the coulomb repulsion at $\hbar\omega_q$. They include therefore in the interaction a repulsive part ($V > 0$) for energies $\epsilon_k > \hbar\omega_q$. The resulting variation of the energy gap at

$0°K$ as a function of $\epsilon_k$ has been shown by Morel and Anderson to have the form represented schematically in Figure 43. Swihart found that a rise of this gap function on moving from the Fermi surface leads to the correct specific heat jump for lead at $T_c$. The relation between calorimetric and magnetic properties indicates that such a gap variation is also consistent with the observed critical field curve for lead and for mercury (Swihart *et al.*, 1965).

A more precise calculation of the energy gap variation with quasi-particle energy $\epsilon_k$ cannot content itself with assigning to the phonons an average energy $\hbar\omega_q$. Instead it must take into account the details of the phonon spectrum as determined, for example, by neutron diffraction (see, e.g., Brockhouse *et al.*, 1962, for lead). It is necessary to consider the different frequency distribution for the longitudinal and transverse phonons. Such a calculation was first carried out numerically by Culler *et al.* (1962), in good agreement with the first experimental indications of energy gap structure in the tunneling experiments of Rowell *et al.* (1962), Giaever *et al.* (1962), and Adler and Rogers (1963). More detailed measurements were later taken by Rowell *et al.* (1963), Bermon and Ginsberg (1964), and Adler *et al.* (1965).

An analytical rather than a numerical treatment of the energy-gap variation due to the phonon spectrum was initiated by Schrieffer *et al.* (1963), and extended to finite temperatures by Scalapino *et al.* (1965). Scalapino and Anderson (1964) discussed the relationship between sign changes in the gap function and so-called Van Hove and other singularities in the phonon spectrum.

In considering an energy gap which changes sign as a function of $\epsilon_k$, it must be remembered that in an experiment involving thermal or electromagnetic absorption across the gap, the quantity actually observed is the energy $E_k$, defined by XI.9. This involves only the square of the gap, and is therefore always positive. However, the details of the variation of the variation of the gap with $\epsilon_k$ can be verified by tunneling experiments, in which the conductance $dI/dV$ is directly proportional to the density of states in the superconductor (Bardeen, 1961a, 1962a; Cohen *et al.* 1962). Schrieffer *et al.* (1963) have however pointed out that for tunneling one cannot use the standard expression for the quasi-particle density of states. This is because when an electron tunnels from one side of the barrier to the other, the initial and final states are not quasi-particle eigenstates of the individual metals making up the tunnel. Instead the appropriate density of states to use is

$$N(E_k) = N(0) \operatorname{Re} \left\{ \frac{E_k}{\sqrt{[E_k^2 - \epsilon^2(0)]}} \right\} \tag{XI.31}$$

in which the energy gap $\epsilon(0)$ is taken to vary with $\epsilon_k$.

The theory of strong coupling superconductors has been summarized by Scalapino *et al.* (1966). Swihart (1966) has extended it to intermediate electron-phonon coupling, so as to be applicable to

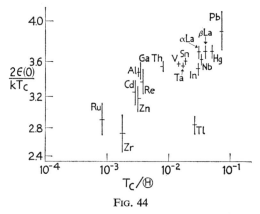

FIG. 44

indium and tin. McMillan and Rowell (1965) have used the theory to deduce the phonon spectrum from tunneling data. More recently, Rowell *et al.* (1965) were in this way able to infer the existence of an impurity band in the phonon spectrum of lead containing small amounts of indium. Adler *et al.* (1966) showed that this band as well as the basic characteristics of the lead phonon spectrum persist even for alloys containing 26 per cent of indium.

A more realistic gap function can yield theoretical justification for the apparent correlation of $\epsilon(0)$ with $T_c/\Theta$. Such a correlation was suggested by Goodman (1958), whose plot of energy gap values against $T_c/\Theta$ for 17 different superconductors is shown in Figure 44.

Goodman used gap values deduced from empirical values of $\gamma$, $H_0$, and $T_c$ by combining XI.13, XI.18, and XI.20, and remembering that $\gamma = \frac{2}{3}\pi^2 k_B^2 N(0)$. This yields

$$\frac{2\epsilon(0)}{k_B T_c} = \frac{4\pi}{\sqrt{3}}\eta^{1/2}, \quad \text{where } \eta \equiv H_0^2/8\pi\gamma T_c^2. \quad \text{(XI.32)}$$

Appropriate values of $2\epsilon(0)/(k_B T_c)$ are listed in Table III.

In general, the stronger the coupling, the larger the value of this ratio. Strong coupling also affects the temperature variation of the energy gap (Swihart *et al.*, 1965). This was verified by Gasparovic *et al.* (1966) and agrees with earlier data of Bermon and Ginsberg (1964).

### 11.8. The specific heat

One can obtain the electronic specific heat in the superconducting phase by twice differentiating with respect to temperature the free energy expression XI.21. At sufficiently low reduced temperatures, for which $2\epsilon(T) \gg k_B T_c$, this yields

$$\frac{C_{es}}{\gamma T_c} = \frac{3}{2\pi^2}\left(\frac{\epsilon(T)}{k_B T_c}\right)^3 \left(\frac{T_c}{T}\right)^2 \left\{3K_1\left(\frac{\epsilon(T)}{k_B T}\right) + K_3\left(\frac{\epsilon(T)}{k_B T}\right)\right\}, \quad \text{(XI.33)}$$

where $K_1$ and $K_3$ are first and third order modified Bessel functions of the second kind. This simplifies in the temperature regions indicated to the following exponential expressions:

$$\left.\begin{aligned}
\frac{C_{es}}{\gamma T_c} &\approx 8\cdot5\exp(-1\cdot44 T_c/T), \quad 2\cdot5 < T_c/T < 6, \\
&\approx 26\exp(-1\cdot62 T_c/T), \quad 7 < T_c/T < 11.
\end{aligned}\right\} \quad \text{(XI.34)}$$

Experimental data at this time exist only in the first of these two regions where they are in good agreement with the BCS values, except for the upward deviation at the lowest temperature which was mentioned earlier (see Figure 29).

Further numerical predictions of the BCS theory include

$$\frac{C_{es}(T_c)}{\gamma T_c} = 2\cdot43. \quad \text{(XI.35)}$$

The following table is taken from [7] (p. 212) and shows how closely

| Element | $\dfrac{C_{es}(T_c)}{\gamma T_c}$ |
|---------|------|
| Lead | 3·65 |
| Mercury | 3·18 |
| Niobium | 3·07 |
| Tin | 2·60 |
| Aluminium | 2·60 |
| Tantalum | 2·58 |
| Vanadium | 2·57 |
| Zinc | 2·25 |
| Thallium | 2·15 |

most experimental values agree with this. The theory also yields that

$$\frac{\gamma T_c^2}{H_0^2} = 0·170, \tag{XI.36}$$

from which one can calculate (see equation II.15) that the predicted coefficient of $t^2$ in the polynomial expansion of the threshold field is

$$a_2 = 1·07. \tag{XI.37}$$

This agrees exactly with the experimental value for tin (VIII.3) and closely with that for several other elements. In terms of the deviation of the threshold field curve from a strictly parabolic variation as displayed in Figure 24, any value of $a_2$ greater than unity corresponds to a curve below the abscissa; only mercury and lead are seen to have deviations corresponding to values of $a_2$ smaller than unity.

The BCS calculations are based on an isotropic model, in which the interaction parameter $V$ does not depend on the direction of $k$ and $k^1$. Pokrovskii (1961) and Pokrovskii and Ryvkin (1962) have investigated the effects of anisotropy on thermal and magnetic properties. They find that in anisotropic superconductors the specific heat ratio in XI.35 should be *smaller* than 2·43, the quantity in XI.36 *larger* than 0·170, and therefore the coefficient $a_2$ *larger* than 1·07. In the second of the papers cited these results are compared with extensive experimental data.

The thermal conductivity in superconductors has been calculated on the basis of the BCS theory for several of the pertinent mechanisms. Bardeen *et al.* (BRT, 1959) and Geilikman (1958) have derived the ratio of the electronic conductivity in the superconducting phase to that in the normal one when this is primarily limited by impurity scattering (equation IX.6). Their results have been well confirmed experimentally, as was discussed in Chapter IX. The derivation of BRT for the case of electronic conduction limited by phonon scattering (equation IX.5) does not, however, lead to the empirical behaviour. Calculations by Kadanoff and Martin (1961) and by Kresin (1959) are in better agreement, but further theoretical work is needed for this conduction mechanism, in which quasi-particle lifetimes due to intermediate and strong coupling effects may again be important (see [7], pp. 272 ff.). Calculations by Tewordt (1962, 1963) indicate that this has little effect in the case of indium and tin, but Ambegaokar and Tewordt (1964) and Ambegaokar and Woo (1965) have successfully applied strong coupling theory to lead and mercury.

BRT as well as Geilikman and Kresin (1958, 1959) have derived the lattice conductivity limited by electron scattering. Experimentally it is very difficult to separate out this part of the heat transport. Where this has been possible (Connolly and Mendelssohn, 1962; Lindenfeld and Rohrer, 1965) the results have been in general agreement with the theoretical predictions.

### 11.9. Coherence properties and ultrasonic attenuation

One of the most striking predictions of the BCS theory arises as a direct consequence of the pairing concept, and experimental verification of this point is thus of particular importance. In a normal metal the scattering of an electron from state $\mathbf{k} \uparrow$ to state $\mathbf{k}' \uparrow$ is entirely independent from the scattering of an electron from $-\mathbf{k} \downarrow$ to $-\mathbf{k}' \downarrow$ or of any other transition. The coherence of the paired electrons in the $\mathbf{k} \uparrow$ and $-\mathbf{k} \downarrow$ states in the superconducting phase, however, makes these two transitions interdependent. The details of the theory (see [7], pp. 212–24) show that the contribution of the two possible transitions interfere either constructively or destructively depending on the type of scattering phenomenon involved. There is constructive interference in the case of electromagnetic interaction, such as the absorption of electromagnetic radiation, and the hyperfine interaction which determines the nuclear relaxation rate. The experimental results expected in these two cases are therefore qualitatively those which follow from a two-fluid model consideration of the total number of electrons available as well as from the density of available states. It has already been mentioned how this explains the observed rise in the nuclear relaxation rate just below the critical temperature (Figure 33).

On the other hand, the contributions of the two transitions interfere destructively in the case of the absorption of phonons, such as occurs in the attenuation of ultrasonic waves. This destructive interference so decreases the probability of absorption that the effect of the increase in density of states on either side of the gap is completely wiped out, and the absorption just below $T_c$ drops very sharply. For low frequency phonons, $\hbar\omega \ll 2\epsilon(0)$, the ratio of attenuation coefficient in the superconducting and normal phase $\alpha_s/\alpha_n$ drops below $T$ with an infinite slope, and is given by

$$\frac{\alpha_s}{\alpha_n} = 2/\{1 + \exp[2\epsilon(T)/k_B T]\}. \qquad (XI.38)$$

This function is shown in Figure 45, which includes experimental points on both tin and indium by Morse and Bohm (1957). It should be contrasted with the theoretical prediction for nuclear relaxation rate, shown in Figure 33.

Within a range of a few millidegrees below $T_c$, the attenuation for transverse waves is found to drop more rapidly than is predicted by the BCS function (Claiborne and Morse, 1964). This is due to the screening by superconducting currents of the fields induced by the sound waves. A treatment by Cullen and Ferrel (1966) for high frequencies has been extended by Fossheim (1967).

Measurements of the ultrasonic attenuation in single crystals of tin in different crystal directions has yielded very convincing demonstration of the anisotropy of the energy gap. When an electron absorbs a phonon, energy and momentum can both be conserved only if the component of the electronic velocity parallel to the direction of sound propagation is equal to the phonon velocity, which is the velocity of sound $S$. Since, however, the Fermi velocity of the electrons, $v_0$, is several orders of magnitude larger than $S$, this is possible only for electrons which move almost at right angles to the direction of sound propagation. Thus a measurement of the attenuation of sound propagated in a particular crystalline direction involves only those electrons whose velocity directions lie in a thin disk at right angles to this direction. The value of the energy gap appearing in equation XI.38 is thus one averaged over this particular disk. Such measurements have been

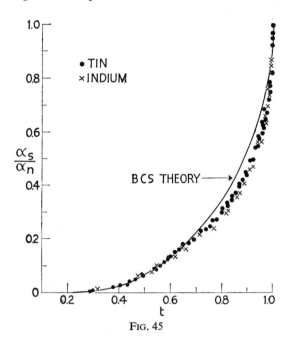

FIG. 45

performed on variously oriented tin single crystals both by Morse *et al.* (1959) and by Bezuglyi *et al.* (1959). Their results are in good agreement and are summarized in the following table:

| *Wave vector* $\mathbf{q}$ | $2\epsilon(0)/k_B T_c$ |
|---|---|
| parallel to [001] | $3{\cdot}2 \pm 0{\cdot}1$ |
| parallel to [110] | $4{\cdot}3 \pm 0{\cdot}2$ |
| perpendicular to [001] and 18° from [100] | $3{\cdot}5 \pm 0{\cdot}1$ |

11

This work has been extended by Bezuglyi *et al.* (1960) and by Leibowitz (1964). Claiborne and Einspruch (1966) have shown that the anisotropy disappears with increasing impurity.

An interesting effect of the energy gap was found by Fagen and Garfunkel (1967) with 9·3 kMhc/s phonons: the direct absorption of phonons by a Cooper pair, with the resulting creation of two quasiparticles. The necessary phonon energy for this is obviously twice the gap. The probability of this effect had been calculated by Privorotskii (1962) and by Bobetic (1964).

The recent interest in gapless superconductors has led to much theoretical work on their dynamical properties, which will be summarized in Section 12.4. Experiments on type II superconductors near $H_{c2}$ — a gapless region — have been reported by Kagiwada *et al.* (1967) and by Gottlieb *et al.* (1967).

Certain anomalies in the attenuation observed for lead (Deaton, 1966) and mercury (Thomas *et al.*, 1966) have been explained as being the result of strong coupling effects (Ambegaokar, 1966; Woo, 1967).

## 11.10. Electromagnetic properties

To describe the many-particle wave function of the superconducting state in the presence of an external field, BCS treat the electromagnetic interaction as a perturbation, and obtain an expansion in terms of the spectrum of excited states in the absence of the field. This wave function is then substituted into an equation of the form III.20 to calculate the current density. Mattis and Bardeen (1958) and also Abrikosov *et al.* (1958) have expanded this to treat fields of arbitrary frequency. The result of the former has been used by Miller (1960) to calculate values of $\sigma_1/\sigma_n$ and of $\sigma_2/\sigma_n$ over a wide range of temperatures and frequencies. His calculations are in excellent agreement with all the experimental results using weak fields at high frequencies, described in Chapter X, if the energy gap is taken as a parameter to be adjusted to its empirical value.

The treatment of a magnetic field as a perturbation in the BCS formulation makes it very difficult to extend it to high field values ($H \approx H_c$). This can be done more readily from a representation of the BCS ideas in terms of Green's functions which has been developed by Gor'kov (1958). A simplified version of this method has been presented by Anderson (1960). The electromagnetic equations occurring in this formulation were shown by Gorkov (1959, 1960) to be equivalent to the Ginzburg-Landau expressions in the region near $T_c$ and under circumstances where $\lambda \gg \xi$. As was pointed out in Chapters V and VII, Gor'kov showed that the energy gap is proportional to the G-L order parameter, so that the dependence of the latter on temperature, magnetic field, and co-ordinates, also applies to the former. The successful application of these ideas to a number of experimental results has been mentioned in Chapter V.

An apparent shortcoming of the original BCS treatment is its lack of gauge invariance. It was suggested by Bardeen (1957) and worked out by various authors that this can be remedied by taking into account the existence of collective excitations. A discussion of this with full references is given in [7] (pp. 252 ff.).

There exists as yet no fully satisfactory explanation that the Knight shift in superconductors does not vanish in any of the elements in which is has thus far been studied: mercury (Reif, 1957), tin (Androes and Knight, 1961), vanadium (Noer and Knight, 1964) and aluminium (Hammond and Kelly, 1964). The Knight shift is defined as the fractional difference in the magnetic resonance frequency of a nucleus in a free ion and the same nucleus in a metallic medium. It is due to the field at the nucleus created by the free electrons, and is usually taken to be proportional to the electronic spin susceptibility. A literal interpretation of the Cooper pairs of opposite spin would lead one to expect that in a superconductor this susceptibility and hence the Knight shift should vanish at $0°K$. A number of authors (see [7], pp. 261–263; Anderson, 1960; Suhl, 1962; Cooper, 1962; Appel, 1965) have suggested why this may actually not be the case, and although none of these explanations appears fully adequate, they have shown that the Knight shift offers no fundamental disagreement with the idea of the BCS theory. A possible trouble spot was removed when careful remeasurements by Hammond and Kelly (1967) on aluminium showed that the shift does extrapolate to zero at $T = 0°K$ in this element.

It is, furthermore, possible that the Knight shift in some of these elements is not primarily due to spin paramagnetism. Clogston *et al.*

(1962, 1964) deduce from the temperature variation of the Knight shift in vanadium that in the superconducting state the dominant contribution due to the d-electron spin does vanish, as the simple theory would predict. This, however, leaves a finite Knight shift due to orbital paramagnetism which involves electrons too far from the Fermi surface to be involved in pairing. Thus this contribution to the Knight shift in vanadium is not affected by the superconducting transition, and perhaps the orbital part is the dominant one in tin and mercury.

### 11.11. Josephson tunneling

The tunneling discussed in Sections 10.6 and 11.7 involves the passage of one or two quasi-particles from one side of the junction to the other. Josephson (1962) predicted additional tunneling currents when both sides of the junction are superconducting, due to the direct passage of coherent Cooper pairs from one side of the barrier to the other.

To understand this one can consider two pieces of superconducting material, initially separated. As the phase of the Cooper pairs is conjugate to their total number (Anderson, 1963), the *absolute* values of the phases in either one of the pieces are arbitrary but their *relative* values are fixed – in fact the phases of all pairs must be the same. This is the precise equivalent of the statement that all pairs have the same momentum. The common pair phase in one super-conducting piece is independent of that in the other, since one can also alter the number of pairs in one without affecting the other. But if now one imagines (Josephson, 1964) the two pieces to be linked by an insulating barrier with a gradually vanishing thickness, the properties of the sandwich system must go over continuously to those of a simple superconductor, with a common phase for all pairs. Hence a current of pairs (whose number is conjugate to their phase) will flow across the barrier so as to equate the phases as soon as the coupling energy becomes comparable to $kT$. This happens when the barrier is very thin ($\sim 10$ Å), and was first observed by Anderson and Rowell (1963).

Because the junction with such a thin barrier behaves like a single superconductor, Anderson (1964) has called this 'weak' superconductivity.

The temperature dependence of this d.c. Josephson effect has been measured by Fiske (1964) following numerical calculations by Ambegaokar and Baratoff (1963).

The magnitude of the Josephson current is given by

$$j = j_1 \sin \Phi. \qquad (XI.39)$$

The coefficient $j_1$ is determined by size, barrier thickness, etc., and $\Phi$ is the phase difference between the two sides of the junction. In the presence of an external magnetic field $H_0$ in the barrier, the current varies spatially along the barrier according to

$$j = j_1 \sin\left(\Phi - \frac{2ed}{\hbar c} H_0 z\right). \qquad (XI.40)$$

$d$ is the barrier thickness, and $z$ the distance along the barrier. Hence, the current at one point may flow in the opposite direction from that at another, and for certain critical field values the total current in fact vanishes. This can be shown to occur whenever the total flux enclosed by the barrier is an integral multiple of the flux quantum $\phi_0$. This effect was first observed by Rowell (1963). The Josephson current will also be affected by the magnetic field which it creates itself, resulting in a self-limitation to certain distances from the barrier edges (Ferrell and Prange, 1963).

The direct Josephson effect described thus far occurs with zero bias voltage across the junction up to a certain maximum value. When this is exceeded, the junction switches to the usual $I$-$V$ curve for quasi-particle tunneling (see Figure 34, Section 10.6). Josephson (1962) predicted that this normal behaviour is accompanied by an alternating supercurrent of frequency $\nu = 2eV/h$. As $2e/h = 483.6$ megacycles/microvolt, the frequency for typical junction voltages of a few millivolts can be as high as $10^8$ cycles/second. Because of this and the very small power levels, this a.c. Josephson effect was first confirmed by indirect evidence (Shapiro, 1963, 1964; Eck *et al.*, 1964; Giaever, 1965) and only quite recently by the direct observation of the accompanying radiation (Yanson *et al.*, 1965; Langenberg *et al.*, 1965, 1966). The former authors reported the observation of

$10^{-14}$ watt. Further experiments have been reported by Silver *et al.*
(1966).

The a.c. Josephson effect can be used for a number of interesting
device applications (Langenberg *et al.*, 1966). In addition to fur-
nishing a simple source of coherent radiation at very high frequencies,
or being used as a sensitive detector or a harmonic multiplier, it can
also provide an extremely sensitive method to measure any of the
quantities appearing in equation XI.40: frequency, voltage, or the
ratio $e/h$.

### 11.12 Macroscopic interference effects

The experiments on the Josephson effect discussed in the preceding
paragraph were carried out on small junctions incorporating a
single 'weak' superconducting link: a thin insulating barrier, or point
contacts made simply by pressing two suitably shaped superconductors together. These already demonstrated fundamental quantum
effects on a macroscopic scale. Even more striking in this respect are a
series of experiments by the Ford group which demonstrate phase
coherence and related effects over distances of over a metre.

These experiments were carried out on a configuration which
essentially consisted of a ring of superconducting material split into
two halves and joined by 'weak' links in two places. Thus superconducting pairs can flow from one side to the other by two routes,
and as their phase is coherent throughout, this results in interference
in precise analogy to an optical interferometer. The essential usefulness of this 'superconducting interferometer' is that, as mentioned
in the preceding section, the phase of the pair current is changed by
external magnetic fields. Whenever the total flux through the ring
changes by one quantum $\phi_0 = hc/2e$, the current phase changes by
$2\pi$. Thus when a varying external field $H$ is applied to the ring, the
superconducting current displays an interference pattern with peak-
to-peak periodicity of $H = (\hbar A/2e)$, where $A$ is the area enclosed by
the ring. With an enclosed area of $\sim 1 \text{ cm}^2$, this gives a periodicity
of $\sim 10^{-7}$ gauss.

The first experiments of this type were reported by Jaklevic *et al.*
(1964) and by Lambe *et al.* (1964), and more recent work has been
summarized by Zimmerman and Silver (1966). The device has been

used as a highly sensitive magnetic flux meter (Lambe *et al.*, 1964), as a way of determining h/m (Zimmerman and Mercereau, 1965) or e/h (Taylor *et al.*, 1966) with great accuracy, and as a superconducting galvanometer responding to magnetic fields generated by minute currents (Clarke, 1966).

# CHAPTER XII

# Superconducting Alloys and Compounds

## 12.1. Introduction

Ever since the discovery of superconductivity there have been many searches for new superconducting materials. Roberts (1961) has recently listed more than 450 alloys and compounds with critical temperatures ranging from $0 \cdot 16°$ up to $18 \cdot 2°$K. In the appearance of superconductivity among these substances there exist certain regularities which were discovered by Matthias (1957) and to which reference was made in Chapter I. One might consider as an ultimate goal of any complete microscopic theory the ability to derive these Matthias rules from first principles. This would be equivalent to being able to calculate with some precision the actual critical temperature of any superconductor. At the moment our understanding of superconducting and of normal metals is still very far from such achievements.

One of the many ways of increasing this understanding is a systematic study of superconducting alloy systems in which solvent or solute are used as controlled parameters. This has been done in a number of experiments.

## 12.2. Dilute solid solutions with non-magnetic impurities

Serin, Lynton, and collaborators (Lynton *et al.*, 1957; Chanin *et al.*, 1959) have investigated the superconducting properties of dilute alloys of various solutes into tin, indium, and aluminium, up to the limit of solid solubility. For low impurity concentrations, of the order of a few tenths of an atomic per cent, $T_c$ decreases linearly with the reciprocal electronic mean free path, independently of the nature of the solute. When plotted against the reduced co-ordinate $\xi_0/l$, where $\xi_0$ is the coherence length of the pure solvent, the fractional change in $T_c$ is the same for elements as different as Sn and Al (Serin, 1960). This is shown in the initial portions of both curves in Figure 46. The existence and the magnitude of this seemingly general effect lend strong support to Anderson's model of impure superconductors (Anderson, 1959). He suggested that the energy gap anisotropy is smoothed out by impurity scattering and disappears when the electronic mean free path is comparable to

$$\frac{\hbar v_0}{2\epsilon(0)} \approx \xi_0.$$

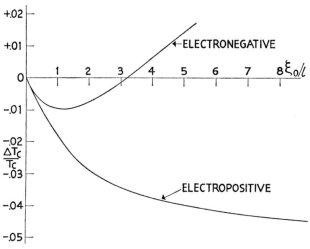

FIG. 46

This should then result in a lowering of $T_c$ by an amount approximately equal to the square of the fractional anisotropy. Nuclear resonance in aluminium (Masuda and Redfield, 1960a, 1962) and ultrasonic and infrared absorption in tin (Morse *et al.*, 1959; Bezuglyi, *et al.*, 1959; Richards, 1961) have shown that the gap in these elements varies by about 10 per cent from its average value, so that $T_c$ should be lowered by about 1 per cent when $l \approx \xi_0$. The measurements of $T_c$ confirm this very well. Recently Caroli *et al.* (1962), Markowitz and Kadanoff (1963), and Tsuneto (1962) have shown in terms of the microscopic theory that Anderson's idea of the smoothing of an anisotropic energy gap indeed leads to a lowering of $T_c$ of the observed magnitude. Hohenberg (1963) has calculated the dependence of $T_c$, the energy gap, and the density of states on the concentration of impurities.

This general mean free path effect on $T_c$ has also been found in tantalum by Budnick (1960). It has been verified by using a number of different ways of scattering the electrons: by mechanical deformation and cold work in aluminium (Joiner and Serin, 1961), by size effects in indium (Lynton and McLachlan, 1962), by quenching (De Sorbo, 1959), electron irradiation (Compton, 1959), neutron bombardment (Blanc *et al.*, 1960), and by using isoelectronic ternary compounds (Wipf and Coles, 1959) in tin.

Figure 46 shows that for $\xi_0/l > 1$, the effect on $T_c$ deviates from the initial linear decrease in a way which depends on whether the solute is electropositive (valence smaller than that of solvent) or electronegative (valence larger). Chiou *et al.* (1961) have extended such measurements to higher concentrations. They found that for both types of impurities $T_c$ ultimately rises to values above that of the solvent, and were able to represent the variation of $T_c$ with impurity concentration in all cases by an empirical relation containing two parameters adjusted according to the particular solvent–solute combination.

According to the BCS theory (equation XI.29), $T_c$ depends on three parameters: an average phonon frequency $\omega_q$ (which is proportional to the Debye temperature $\Theta$), the density of normal electron states at the Fermi surface, $N(0)$ (which is proportional to the Sommerfeld $\gamma$), and the BCS interaction parameter $V$. Specific heat measurements on tin alloys have recently enabled Gayley *et al.* (1962) to find the effects of the addition of indium, bismuth, and indium antimonide on the values of $\gamma$ and of $\Theta$ for tin. One can use equation XI.29 to calculate the corresponding change in $T_c$ (Ginsberg, 1964; Gayley, 1964). This seems to account for most of the difference in the behaviour of electropositive and electronegative solutes, at least in the case of indium and bismuth, but not for the increase in $T_c$ at high solute concentrations. One concludes that this increase is mainly due to effects of alloying on the interaction energy $V$ (Ginsberg, 1965).

Any attempt to calculate $V$ in the presence of impurities has to take into account that with scattering the wave vectors **k** are no longer good quantum numbers. Hence the question arises of the criterion for pairing of the electrons. Abrahams and Weiss (1959) and Anderson (1959) have pointed out that in impure superconductors Cooper pairs are formed of two electrons the wave functions of which are identical except for the reversal of the time co-ordinate, and which have the same energy. The former authors have used this to deduce

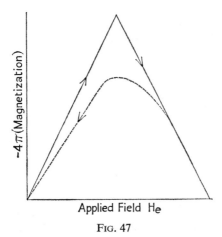

FIG. 47

several impurity effects, the magnitude of which is difficult to estimate. Anderson (1959, 1960) has discussed the general implications of the use of time-reversed wave-functions. Detailed microscopic calculations of the impurity effects on the superconducting parameters have been attempted by Caroli *et al.* (1962), Markowitz and Kadanoff (1963), Brink and Zuckerman (1965), Zuckerman (1965), and Appel (1967).

Merriam and co-workers (see, e.g., Merriam, 1966; Merriam *et al.*, 1967) have carried out extensive investigations of the critical temperature of alloy systems over wide ranges of concentration. This has yielded information about the alloy constitution diagrams, their electronic structure, Brillouin zone overlap, and order-disorder transformations.

It is interesting to note that the work on carefully homogenized and annealed solid solutions has shown these to be 'well-behaved' super-conductors according to several criteria. Transitions occur within a few millidegrees, and very little flux remains in suitably oriented cylindrical samples after an external field has been removed (Budnick *et al.*, 1956). Also the absorption edge of infrared radiation at the gap frequency can be very sharp (Leslie and Ginsberg, 1962). Detailed magnetization curves (Lynton and Serin, 1958) however, show that the transitions for such alloys are nevertheless not fully reversible, as shown in Figure 47 for 3·11 per cent In-Sn cylinders transverse to an external field. In decreasing field the magnetic moment does not attain its full diamagnetic value until $H$ vanishes. This indicates that flux is initially trapped, but then leaks out as suggested by Faber and Pippard (1955b).

The impurity effects described thus far occur with light, non-magnetic impurities. When heavy, transition metal impurities are dissolved, this can lead to localized magnetic states for some solvents, as will be discussed in the next section. Even when there are no localized moments, as e.g. in alloys of aluminium with ferromagnetic transition elements (Cr, Mn, Fe, Co, and Ni), the effect of impurities is a drastic lowering of the critical temperature (Boato *et al.*, 1963, 1964, 1966; Aoki and Ohtsuka, 1965). Ratto and Blandin (1967) have studied this theoretically.

### 12.3. Compounds with magnetic impurities

Matthias and collaborators have traced the occurrence of super-conductivity in a large number of compounds containing para-magnetic and ferromagnetic impurities (Matthias, 1960). Their results can be summarized as follows:

Ferromagnetic transition elements with $3d$ electrons (Cr, Mn, Fe, Co, and Ni) put into fourth column superconductors (Ti, Zr) raise $T_c$ more than do corresponding amounts of transition elements with $4d$ electrons (Re, Rh, Ru, etc.) (Matthias and Corenzwit, 1955; Matthias *et al.*, 1959b). At the same time magnetic measurements on Ti-Fe and Ti-Co alloys indicated the absence of localized moments. The effect of the $4d$ electrons can be attributed to the increase in the number of valence electrons per atom toward five, a number particularly favourable for superconductivity. The extra rise with 3$d$ electrons is attributed to a magnetic electron–electron interaction favouring superconductivity. For the same reason adding Fe (3$d$ electrons) to a $Ti_{0.6}V_{0.4}$ compound lowers $T_c$ *less* than does an equal amount of Ru (4$d$ electrons): in both cases $T_c$ is decreased because the number of valence electrons per atom rises beyond five, but with Fe the apparent magnetic interaction counteracts this in part. It must be pointed out, however, that ferromagnetic transition elements with 3$d$ electrons put into fifth column superconductors (Nb, V) lower $T_c$ in approximate agreement with the expected effect due to the change in valence electrons per atom (Müller, 1959). There does not appear to be any added effect due to the magnetic nature of the impurities. Why such effects should appear with fourth column metals but not with fifth column ones is far from clear, as in neither case are there any localized magnetic moments associated with the 3$d$ solute atom.

More recently Cape (1963) has measured the electrical and magnetic properties of very carefully prepared alloys of Ti containing 0.2 to 4 at % Mn. Depending on the method of preparation these specimens are either in a single hexagonal close packed (hcp) phase, or contain an admixture of a second, body centred cubic (bcc) phase. Localized moments exist only in the hcp phase, which however is *not* superconducting. This is consistent with the usual suppression of superconductivity by impurities retaining localized moments (see below). The non-magnetic bcc phase, on the other hand, has a transition temperature which is raised above that for pure Ti by an amount commensurate with the increase in the number of valence electrons. Hake *et al.* (1962) had earlier deduced from their measurements of transport properties that the hcp phase of Ti–Cr, Ti–Fe, and Ti–Co also carried localized magnetic moments. In addition there is calorimetric evidence (Cape and Hake, 1963) that in Ti–Fe samples only a small fraction of the volume is superconducting. These results throw considerable doubt on Matthias' speculation that iron-group impurities which do *not* carry a localized magnetic moment enhance superconductivity by means of a magnetic interaction between electrons.

While 3$d$ impurities in fifth column metals (for example Nb) do not show any evidence for a localized moment, they do when put into sixth column metals (for example, Mo), and in fact Matthias *et al.* (1960) found that the change in behaviour occurs in Nb-Mo solutions at a concentration of about 60 per cent Mo. One would therefore expect some special effects on $T_c$ to appear in $3d$ compounds with sixth column metals. Until the recent discovery of the superconductivity of Mo, no such metal was known to be superconducting. For that reason, this effect was studied on superconducting $Mo_{0.8}Re_{0.2}$, and indeed small amounts of $3d$ impurities lower $T_c$ far more than one

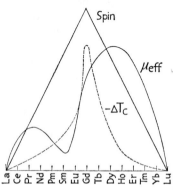

FIG. 48

would expect from valence effects. It is in fact this which made the discovery of the superconductivity of Mo so difficult: a few parts per million of iron are enough to depress $T_c$ below the measurable range (Geballe *et al.*, 1962). A less abrupt decrease in $T_c$ is obtained when rare earth elements with $4f$ electrons are put into lanthanum (Matthias *et al.*, 1958b, 1959a). The magnitude of this decrease, for each per cent of rare earth impurity, is correlated with the spin rather than with the effective magnetic moment of the solute. This is shown in Figure 48 in which $-\Delta T_c$ for each per cent, the spin, and the effective moment $\mu_{eff}$ are plotted for the different rare earths. A higher effective moment, in fact, appears to tend to raise $T_c$, perhaps for the same reason as in the case of the $3d$ impurities in fourth column metals: erbium, with spin $3/2$ and large moment lowers $T_c$ *less* than does an equal percentage of neodynium, which has the same spin but a smaller moment.

All these compounds containing $4f$ electrons show ferromagnetic behaviour at somewhat higher concentrations of the rare earth solutes, with the Curie temperature rising with increasing number of $4f$ electrons. Such dilute ferromagnetism has not been observed for compounds with $3d$ electrons which indicates that the $s$–$f$ magnetic interaction is rather long range, while the $d$–$d$ one is a short-range inter-action effective only through nearest neighbours, which is impossible in dilute solutions (Matthias, 1960).

Interesting analogies in the variation of the Curie temperature and the superconducting critical temperature are found by investigating the magnetic characteristics of so-called Laver phases $AB_2$, where $B$ is germanium or a noble metal (Ru, Os, Ir, Pt) and $A$ is either a rare earth with $4f$ electrons $(A')$ or one of the group Y, Sc, Lu, or La $(A'')$, none of which contain $4f$ electrons (Suhl *et al.*, 1959a). $A'B_2$ is always ferromagnetic, $A''B_2$ always superconducting. Comparing the Curie temperatures of the former with the critical temperatures of the latter one finds a similar dependence on spin and on the number of valence electrons per atom. This is but one of a number of interesting correspondences which Matthias has found between superconductivity and ferromagnetism. There are, for example, several groups of isomorphous compounds which are either superconducting or ferromagnetic (see, for example, Matthias *et al.*, 1958a; Compton and Matthias, 1959; Matthias *et al.*, 1962). Also, the appearance of localized moments when a ferromagnetic impurity is put into a non-magnetic transition element seems to depend on the number of valence electrons in a manner similar to the criterion for the appearance of superconductivity (Matthias, 1962). Matthias has frequently suggested that an electron configuration favourable to superconductivity may also be favourable to ferromagnetism.

The possible coexistence of superconductivity and ferromagnetism in the same substance has been investigated in lanthanum-rare earth binary compounds (Matthias *et al.*, 1958b) and in Laver phase mixtures $(A'_{1-x}A''_x)B^2$ (Matthias *et al.*, 1958c; Suhl *et al.*,1959a). Susceptibility measurements (Bozorth *et al.*, 1960) as well as calorimetric ones (Phillips and Matthias, 1961; Finnemore *et al.* 1965a) indicate that ferromagnetism and superconductivity occur in the same sample, and the more recent data in particular suggest that these two types of magnetic behaviour do not merely exist side by side in different portions of the specimen, but indeed coexist on a microscopic scale. Anderson and Suhl (1959) showed that this would be energetically possible if the ferromagnetic alignment occurs over a very small range, of the order of 10 Å. They call this 'cryptoferromagnetism.'

Since then Gor'kov and Rusinov (1964) have developed a detailed theory for the phenomenon. They showed that the scattering by the magnetic impurities causes one. of the Cooper-pair spins to flip so as to give the superconductors a finite spin susceptibility even at $0°K$. This is the mechanism which had earlier been suggested to explain the non-vanishing Knight shift (see Section 11.10). The finite spin susceptibility in turn facilitates ferromagnetic ordering in the superconducting state at sufficiently low temperatures.

In extending the Gor'kov-Rusinov theory, Fulde and Maki (1966a) showed its formal equivalence to the case of magnetic impurities with randomly oriented spins. They also derived the temperature variation of the critical field $H_{c2}$ and of $\kappa_2(T)$ (see Section 7.4), obtaining good agreement with the measurements of Finnemore *et al.* (1965b) on Gd-La alloys. Benneman and Garland (1967) calculated the specific heat of such alloys, and obtained results which agree with the data of Phillips and Matthias (1961) and of Finnemore *et al.* (1965a).

Suhl and Matthias (1959) have treated the general problem of the lowering of $T_c$ due to the presence of magnetic impurities by extending an argument of Herring (1958), according to which the polarization due to the coupling of the conduction electrons with the spins of the paramagnetic impurity ions lowers the free energy in both the normal and in the superconducting phases. The free energy is lowered by each electron-spin scattering interaction by an amount proportional to the reciprocal energy difference between the initial and final electron state. In the normal state this difference can be arbitrarily small, in the superconducting case this difference cannot be smaller than the energy gap. As a result the free energy of the normal phase is lowered more than that of the superconducting one, and the onset of superconductivity therefore occurs at a lower temperature. Suhl and Matthias ignore the small changes in the interaction matrix element $V$, and as a result their prediction ($\partial T_c/\partial c \to 0$) for very small magnetic impurity concentrations is probably wrong. Abrikosov and Gor'kov (1960) show that magnetic impurity effects on $V$ initially lowers $T_c$ linearly with impurity concentration. At much higher concentrations $\partial T_c/\partial c \to \infty$ (Suhl and Matthias, 1959; Baltensperger, 1959; Suhl, 1962). This is in agreement with the data of Hein *et al.* (1959) and of Crow and Parks (1966). Extensions of the Abrikosov-Gor'kov theory to higher-order approximations (Liu, 1965; Griffin, 1965; Maki, 1967a) do not much change the calculated shift of critical temperature with magnetic impurity concentration.

## 12.4. Gapless superconductivity

Abrikosov and Gor'kov (1960) as well as De Gennes and Sarma (1963) derived that magnetic impurities lower the energy gap more rapidly than the transition temperature. There should thus be a range of concentration for which the alloy has a finite critical temperature at which its d.c. resistance disappears without the existence of an energy gap. Indeed Woolf and Reif (1965) have found this paradoxical behaviour to exist. They measured the electrical resistance as well as the tunneling characteristics of a number of lead and indium film containing magnetic impurities. The gap decreased twice as rapidly as the transition temperature, and an indium film containing 1 at $\%$ Fe, for example, had no resistance below $3°K$ but a perfectly ohmic tunneling conductance. Further experimental evidence is provided by the specific heat measurements of Finnemore *et al.* (1965a).

Skalski *et al.* (1964) carried out the first complete calculation of the density of states and of the electrodynamic properties of gapless superconductors. One can see from their results that this kind of superconductivity does not violate any fundamental principle. Instead of there being a finite range of energy in which there is a total absence of states, the 'gapless' superconductor has in such a range merely a very low density of states. This continues to lead to a real conductivity $\sigma_1$ which is at all frequencies less than the normal conductivity $\sigma_N$ (see Figure 38), and according to the Ferrel-Glover

12

sum rule (see Section 10.8) this must be compensated by a delta function conductivity for zero frequency, which is the criterion for superconductivity. The phenomenon of gapless superconductivity has attracted much theoretical and experimental attention during recent years because it has been found to occur quite frequently. Whenever the superconducting state is strongly perturbed – by localized magnetic moments as discussed above, by an external magnetic field, or by spatial variations of the order parameter – and the perturbation is a pair-breaking mechanism which turns the normal state transition into a second-order one, the superconductor will be gapless in the neighbourhood of the transition. As a result, gapless superconductivity occurs in the following situations in addition to the one already cited for alloys with magnetic impurities:

(a) thin films and other small specimens in the presence of an external field (Maki, 1963b, 1964d; DeGennes and Tinkham, 1964),

(b) thin films carrying a near-critical current (DeGennes and Tinkham, 1964; Fulde, 1965),

(c) thin films in a perpendicular magnetic field (Maki, 1965; Lasher, 1967),

(d) type II superconductors near $H_{c2}$ (Maki, 1964a; DeGennes, 1964),

(e) superimposed superconducting and normal films (Fulde and Maki, 1965, 1966; DeGennes and Mauro, 1965),

(f) superconducting surface sheaths near $H_{c3}$ (Maki, 1966).

The clearest evidence for the absence of the gap can of course be obtained by tunneling experiments, which provide a direct sampling of the density of states as a function of energy. Such experiments have been carried out on type II superconductors (Guyon *et al.*, 1965; Guyon, 1966; Orsay Group, 1966), on thin films in parallel and perpendicular magnetic fields (Guyon *et al.*, 1967), and on superimposed films (Guyon *et al.*, 1966; Claeson *et al.*, 1967; Hauser, 1966; Adkins and Kingston, 1966).

The transport properties of gapless superconductors have been subject to extensive calculations and measurements, especially in the dirty limit. The electronic part of the thermal conductivity in thin films was treated by Maki (1964d) and that in alloys containing magnetic impurities by Ambegaokar and Griffin (1965). Caroli and Cyrot (1965) have calculated the thermal transport for cases where the order parameter is position dependent, such as in type II superconductors near $H_{c2}$ and in surface sheaths near $H_{c3}$. Their results have been confirmed by the experiments of Dubeck *et al.* (1964) and Lindenfeld *et al.* (1966) on type II specimens, and of Mochel and Parks (1966) and of Smith and Ginsberg (1968) on the surface sheath.

Kadanoff and Falko (1964) have calculated the ultrasonic attenuation in alloys with magnetic impurities, and Maki (1966b) that for type II and surface-sheath situations. Measurements of Gottlieb *et al.* (1967) in the mixed state are in substantial agreement.

The calculations for gapless superconductivity in *pure* materials differs strikingly from the dirty case (Orsay Group, 1966; Juranek *et al.*, 1966; Cyrot and Maki, 1967). Detailed calculations of the thermal conductivity in pure gapless superconductors have been made by Maki (1967b) and on ultrasonic attenuation by Caroli and Matricon (1965), Cooper *et al.* (1965), and Maki (1967a). The latter results have been confirmed by the measurements of Kagiwada *et al.* (1967) and of Gottlieb *et al.* (1967).

## 12.5. Superimposed metals

A recent series of experiments by Meissner (see Meissner, 1960 for full references) has revived interest in the question whether thin layers of superconducting material deposited on a normal metal would themselves become normal, and whether conversely sufficiently thin layers of normally non-superconducting metal would become superconducting when in contact with a superconductor. Such superimposed metals differ from the sandwiches used in the tunnelling experiments by the absence of an insulating layer.

Cooper (1961) has suggested an intuitive argument for the possibility that the superconducting properties of thin metallic films may be strongly affected by direct contact with other metals. He emphasized that in the BSC theory one must clearly distinguish between the range of the attractive interaction between electrons, and the distance over which as a result of this interaction the electrons are correlated into Cooper pairs. The range of the interaction is very short ($10^{-8}$ cm); the 'size' of the wave packets of the pairs, on the other hand, is of the order of the coherence length, that is, $10^{-4}$ cm. This, as Cooper points out, is analogous to the difference between the range of the nuclear interaction and the much larger size of the resulting deuteron wave packet. Because of this long coherence length the Cooper pairs can extend a considerable distance into a region in which the interaction between electrons is not attractive. Thus when a thin layer of super-conducting material is in contact with a layer of normal metal, the zero-momentum pairs formed because of the attractive interaction in the superconductor extend into both layers. As a result the ground state energy of this thin bimetallic layer is characterized by some aver-age over both metals of the parameter $N(0)\,V$, which in turn deter-mines the energy gap of the layer and its transition temperature, according to equations XI.18 and XI.29. The form of this average of course depends on the nature of the boundary between the two metals; the better the contact, the more effective is a superimposed layer in changing the properties of the substrate. Regardless of how one accounts for this, one would expect the average to depend also in some manner on the relative thickness of the two layers. The thicker the normal layer, the smaller the average interaction, and the more the energy gap width and the transition temperature are decreased from the values they would have if only the superconductor were present. Similarly a combination of two superconductors would be expected to have a $T_c$ somewhere between the $T_c$ values of the two materials, varying from one extreme to another as the relative thickness of the two layers is varied. DeGennes (1964) has shown that for two thin layers of thickness $d_N$ and $d_S$ respectively, the effective value of $N(0)\,V$ to be used in calculations is given by

$$[N(0)\,V]_{\text{eff}} = \frac{N_N^2\,V_N d_N + N_s^2\,V_s d_s}{N_N d_N + N_s d_s}\,. \qquad \text{(XII.1)}$$

$N$ and $V$ are the appropriate density of states and interaction para-meter in the two metals.

These qualitative conclusions presuppose that both layers of the bimetallic film are sufficiently thin so that the coherent electron pairs extend over the entire volume. One expects the critical thickness for this to be of the order of the coherence length, although it is not clear whether this should be the ideal value $\xi_0$, or the mean free path limited value $\xi(l)$. If one of the two superimposed metals is much thicker than whatever critical length is appropriate, then presumably the average interaction is determined by the ratio of the smaller thickness to the critical length (De Gennes and Guyon, 1962; Orsay Group, 1966).

The detailed theory of what has become known as the *proximity effect* was worked out by DeGennes and co-workers (DeGennes and Guyon, 1962; DeGennes, 1964) and extended by Werthamer (Werthamer, 1963b; Hauser *et al.*, 1964) and Moorman (1967). Early experimental studies (Misener and Wilhelm, 1935; Meissner, 1960; Smith *et al.*, 1961; Hilsch and Hilsch, 1961) indicated qualitative agreement with the expected decrease of the critical temperature with increasing normal metal thickness, but quantitative results for a long time appeared erratic and questionable. Rose-Innes and Serin (1961) showed how strongly data can be influenced by the interpenetration of the two metals, even under conditions which quite preclude ordinary bulk intermetallic diffusion, and DeGennes (1964) suggested that improper evaporation procedures could easily produce a thin oxide layer between the two metals.

During recent years better experimental techniques have led to systematic and reproducible results which are in good agreement with the theory. Hauser and Theuerer (1965) and Bergmann (1965) studied cases where the 'normal' metal was actually a superconductor with a low critical temperature; Hauser *et al.* (1964) and Von Minnigerode (1966) layers with metals such as copper which are not known to be superconducting. The results suggest the possibility of superconducting copper at very low temperatures. The proximity effect with normal metals containing dilute magnetic impurities was measured by Hauser *et al.* (1966). They also looked at the effect of ferromagnetic layers, as did Groff and Parks (1966).

Tunneling studies of the proximity effect at very low temperatures (Adkins and Kingston, 1966; MacMillan, 1966) can yield information about the sign of the interaction parameter $V$. Again there are indications that in copper it is positive, i.e. attractive, suggesting that copper may become superconducting. Tunneling measurements near the critical temperature of the double layer (Claeson and Gygax, 1966; Guyon *et al.* 1966; Burger and Martinet, 1966) generally agree with the theoretical treatment of Fulde and Maki (1965, 1966) and DeGennes and Mauro (1965). Surface impedance studies by Fanelli and Meissner (1966) and by Waldram (1966) also agree with theory.

Extensive observations of the proximity effect in the presence of a magnetic field have been carried out, mainly by the Orsay Group, and have been summarized by Deutscher and DeGennes (1968).

*Note added in proof*: Recent thermal conductivity measurements of the proximity effect suggest that near the normal-superconducting boundary a region of the specimen remains gapless at all temperatures. (G. Deutscher, P. Lindenfeld, and R. D. McConnell, *Phys. Rev. Lett.* **21**, 79, 1968.)

# CHAPTER XIII

# Superconducting Devices

## 13.1. Research devices

The characteristics of superconductors have for a long time already been put to use in many low temperature experiments. It is very common to use niobium wires, for example, in electrical connections to samples which one wishes to isolate thermally as well as possible. Such wires are superconducting with a high critical field throughout the entire liquid helium temperature range, and combine low thermal transport with perfect electrical conductivity. The use of lead wires as heat switches at temperatures below 0·1°K has been mentioned in Chapter IX.

More specialized research devices using superconducting components of varying complexity have been suggested or used frequently, and it is possible in this cursory survey to mention only a few of these. A number of such devices have been developed to detect very small potential differences, as occur, for instance, in studies of thermoelectric powers. Pippard and Pullan (1952) improved earlier designs by Grayson Smith and co-workers (Grayson Smith and Tarr, 1935; Grayson Smith *et al.*, 1936) by using a single turn of superconducting wire to construct a galvanometer capable of detecting e.m.f.s of $10^{-12}$ V. With a resistance as low as $10^{-7}$ ohm this required a current sensitivity of only $10^{-5}$ amp; the time constant $L/R$ was kept short by the single turn design which reduced the effective inductance. A superconducting magnetic shield made possible controlling fields as low as 0·01 gauss. Clarke (1966) describes a way of using Josephson tunneling in a superconducting galvanometer.

A different approach to the measurement of very small potentials was suggested by Templeton (1955b) and by De Vroomen (De Vroomen, 1955; De Vroomen and Van Baarle, 1957). These authors designed 'chopper' amplifiers in which the small d.c. signal is converted into an alternating one by passing through a superconducting

173

wire which is modulated into and out of the normal state by being placed in an alternating magnetic field. The resulting oscillating potential across the wire is then amplified in a conventional manner. These devices can operate stably with a noise level at about $10^{-11}$ V. A somewhat simpler version of such a modulator-amplifier has been designed by Kachinskii (1965). Templeton (1955a) has also designed a superconducting reversing switch to suppress undesirable thermal voltages in measurements of potential differences of the order of about $10^{-6}$ V.

Many low temperature experiments as well as superconducting magnets require rather high direct currents at very low voltages. To avoid the use of thick electrical leads which would bring too much heat into the helium dewar, Olsen (1958) has designed a superconducting rectifier and amplifier which, together with a low temperature transformer, allows one to feed in a low alternating current through thin leads. The rectification occurs as the current flows through a superconducting wire placed in an external, nearly critical field, such that the field due to the current in one direction is sufficient to make the wire normal during about one-half of each cycle. Such a system has recently been improved by Buchold (1964).

An alternative to using a strong current for high field generation is to use a magnetic 'flux pump' – a device which creates a high flux density by an irreversible motion and compression of flux lines. A complete review of several such devices has been given by Van Beelen *et al.* (1965).

D. H. Andrews *et al.* (1946) made use of the change in resistivity at the superconducting transition in designing a bolometer. More recent versions of such devices have been described by Franzen (1963), Low and Hoffman (1963), and Lalevic (1966). A different superconducting radiation detector was suggested by Burstein *et al.* (1961) who pointed out that a suitably biased tunneling junction would respond to absorption of electromagnetic radiation in the microwave and submillimetre range. Shapiro and Janus (1964) were the first to demonstrate this photon-assisted tunneling, which was discussed in Section 10.6. The use of Josephson junctions both as generators and as detectors of very high frequency radiation has been discussed by Langenberg *et al.* (1966).

For work at high frequencies superconducting metals may also be used to construct or to coat resonant cavities of extremely high $Q$. This has been discussed by Maxwell (1960). Following preliminary experiments by Fairbank *et al.* (1964) as well as by Ruefenacht and Rinderer (1964), Schwettman *et al.* (1965) have more recently reported reaching values of $Q$ well above $10^9$ by using lead-plated copper cavities. These could be used in high energy linear particle accelerators (Parkinson, 1962; Fairbank *et al.* 1964), and a prototype is under development at Stanford University.

Many of the devices mentioned in this section have been discussed by Parkinson (1964). A complete bibliography of superconducting devices and of related literature has been compiled by Goree and Edelsack (1967).

## 13.2. Superconducting magnets

As early as 1931, De Haas and Voogd found critical fields as high as 15 kgauss in some lead–bismuth alloy wire. Other instances of relatively large values of the critical field have been observed for many alloys and for strained or impure samples of the superconducting elements. For niobium published values of the critical field at $0°K$ vary from about 1950 to 8200 gauss. Quite recently Kunzler *et al.* (1961b) discovered $Nb_sSn$ to have a critical field of about 200 kgauss, and similar critical fields have since been found in other substances. These seem to be either intermetallic compounds of the $\beta$-wolfram structure, or body centered cubic alloys. When suitably prepared these materials remain superconducting while carrying current densities as high as $5 \times 10^4$ amp/cm$^2$ in fields almost up to the critical value. Goodman (1961) was the first to suggest that these high field substances are type II superconductors. This was quickly confirmed by a number of experimental results. The specific heat values of $V_3Ga$ observed by Morin *et al.* (1962) are consistent with type II behaviour (Goodman, 1963b), as are the critical fields observed by Berlincourt and Hake (1962, 1963) in the low current limit for many high field alloys and compounds. Hauser (1962) and Swartz (1962) further showed that the magnetization curves of suitably prepared specimens are those of type II materials. A systematic study of the role of defects on the distortion of these curves was carried out by Livingston (1963, 1964).

However, Gorter (1962a, b) has pointed out that a homogeneous superconductor with uniform negative surface energy cannot in the presence of a transverse magnetic field carry the high current densities which are actually observed in most of the compounds and alloys under discussion. This can best be understood in terms of the vortex structure of the mixed state which is created by the external field (see Section 6.6). When a current passes through the specimen at right angles to the vortices, it interacts with the latter so as to push them out of the specimen. This, as was mentioned in Section 6.6, can be prevented only if the vortices are pinned down by local variation of the surface energy, as would be present if the specimen were inhomogeneous. Indeed, there is much evidence that the high current carrying capacity is associated with the presence of dislocation in cold-worked specimens (Hauser and Buehler, 1962). Annealed samples may still have a very high critical field while carrying a low current density, but turn normal when the latter is increased. Rose-Innes and Heaton (1963) have used Ta–Nb wire to show very strikingly how sample treatment can change the current carrying capacity without changing the critical field.

Thus the present picture of high field superconductors is that basically they are materials characterized by a negative surface energy. They are further able to carry high current densities in high fields if through cold work they are made to contain a high density of dislocations which pin down the current carrying regions. A nearly uniform distribution of these dislocations explains why the critical current increases as the cross-sectional area of the specimen (Lock, 1961a; Hauser and Buehler, 1962).

The ability of some superconductors to carry high current densities in high fields, of course, suggests their use in the winding of magnets. Yntema (1955) described a superconducting solenoid wound with niobium wire and producing up to 7 kgauss, but this received little attention. In 1960 Autler wound a niobium solenoid creating a field of 4·3 kgauss, and since then the interest in the subject has grown explosively, with much scientific and technical activity in a large number of laboratories. Kunzler *et al.* (1961a) and others used $Mo_3Re$

to wind solenoids producing up to 15 kgauss; much higher fields were achieved soon thereafter as a result of work with $Nb_3Sn$ (Kunzler *et al.* 1961b), $Nb_2Zr$ (Kunzler, 1961; Berlincourt *et al.*, 1961) and NbTi (Coffey *et al.*, 1964). Solenoids wound of these materials have produced fields in excess of 100 kgauss, and both suitable superconducting wire as well as entire solenoid assemblies have become commercially available. A recent review of progress in this rapidly developing field has been given by Laverick (1967), while earlier references can be found in [11], Kropschot and Arp (1961), and Berlincourt (1963).

### 13.3. Superconducting computer elements

Much research and development work is currently being devoted to attempts to use superconductors both as switching devices and as memory storage elements in electronic computers. The basic idea for a superconducting switching element originated with Buck (1956) who invented the cryotron. This consists of a layer of thin (0·003 in.) niobium wire wound on to a thicker (0·009 in.) tantalum wire. A sufficiently large current through the former, called the control winding, can quench the superconductivity of the latter, called the gate. The two materials are chosen because the convenient operating temperature of 4·2°K is only a little below the critical temperature of Ta, but much lower than that of Nb, so that a control current sufficient to 'open the gate' is still much less than the critical current of the control. The diameter of the gate is furthermore kept large so as to maximize the amount of gate current, $I_g$, which can be controlled by the control current, $I_c$. Calling $H_c$ the critical field of the tantalum gate at the operating temperature, and $D$ its diameter, then

$$(I_g)_{max} = H_c \pi D, \qquad (XIII.1)$$

and

$$I_c = \frac{H_c}{n}, \qquad (XIII.2)$$

where $n$ = number of turns/unit length of control winding. Thus

$$\frac{(I_g)_{max}}{I_c} = \pi D n. \qquad (XIII.3)$$

This is the 'gain' of the cryotron, which must be kept at a value greater than unity in order that the gate current of one cryotron can be used to control another.

A great variety of logical circuits can be built up by making use of this reciprocal control of a number of cryotrons. Most of these circuits contain the basic flip-flop or bistable element, shown in Figure 49. Current through this element can flow in either one or the other branch and, once established in one, will flow in it indefinitely since it makes the other one resistive. The choice of branch can be dictated by placing a further cryotron gate in series with each branch, and controlling this by an outside signal, which can 'open the gate', making

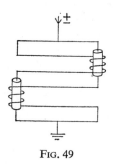

FIG. 49

the corresponding branch resistive and forcing the current into the other path. This is shown in Figure 50, which also indicates that if each branch also controls the gate of a read-out cryotron, the position of the bi-stable element can be read. Figure 51 shows other basic logical circuits using cryotrons; the current through the heavy line flows only if: (*a*) cryotron *A or B* is open, (*b*) cryotrons *A and B* are open, (*c*) *neither A nor B* are open. More complicated logical circuits are discussed by Buck (1956) as well as in review articles by Young (1959), by Haynes (1960), and by Lock (1961b).

Basically all these cryotron circuits consist of a number of parallel superconducting paths between which the current can be switched by the insertion of a resistance into the non-desired branches. Under steady-state conditions the power dissipation is zero as long as there is always at least one path which remains superconducting. The speed with which the resistance can be inserted, that is, the speed with which a given gate can be made normal, depends on the basic phase transition time and is small enough ($\approx 10^{-10}$ sec) not to be a limiting factor at this time (see, for instance, Nethercot, 1961; Feucht and Woodford, 1961). On the other hand, the switching time from one current path to another is determined by the ratio $L/R$, where $L$ is the inductance of the superconducting loop made up of the current paths, and $R$ the resistance introduced by an opened gate. The usefulness of wire-wound cryotrons is severely limited by the fact that this time is no less than $10^{-5}$ sec, even if the gate consists of a tantalum film evaporated

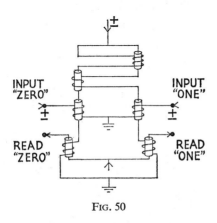

INPUT "ZERO"     INPUT "ONE"

READ "ZERO"     READ "ONE"

Fig. 50

on to an insulating cylinder. Because of this all current research and development effort is directed toward making thin film cryotrons consisting of crossed or parallel gate and control films separated by insulating layers, and placed between additional superconducting shielding films called ground planes. The resistance of the thin film gates is comparable to that of a wire gate, but the ground planes confine magnetic flux to a very small region and thus result in $L/R$ values of the order of $10^{-8}$–$10^{-10}$ sec. Cryogenic loops with a time constant of $2 \times 10^{-9}$ sec have been operated (Ittner, 1960b). An account of many of the design considerations governing such thin film cryotrons can be found in several papers in [9].

Suggestions for superconducting memory devices were advanced simultaneously by Buckingham (1958), Crittenden (1958), and Crowe (1958). Their devices are basically quite similar and make use of the fact that a current induced in a superconducting ring will persist indefinitely. Since the current can circulate either way one has the possibility of a two-state memory storing one bit of information with no dissipation of power other than that required to maintain the low temperature. Of the three suggestions it is that of Crowe on which in recent years most attention has been concentrated and which will be briefly described here. Before doing so it might be noted that persistent current memory devices have in common with switching

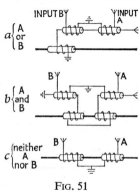

Fig. 51

cryotrons that a current in one superconducting circuit quenches the superconductivity in another. There is, however, no need for a greater-than-unity gain, as the controlled current is not in turn used to drive another unit. One therefore often calls the memory elements low gain cryotrons.

The Crowe cell basically consists of a thin film of superconducting material (for example, lead) with a small hole, a few millimetres in diameter, which has a narrow cross-bar running across it. This is shown schematically in Figure 52. A drive 'wire' in the form of a second narrow strip lies just above the cross-bar, separated only by a thin insulating layer. As long as the entire configuration remains superconducting, the magnetic flux threading the hole must retain its original value, which we shall take to be zero. Therefore if a current is passed through the drive wire, it will induce currents in the cross-bar and the remainder of the film. The direction of this induced circulating current will be such as to keep the flux from penetrating, and results in a flux distribution indicated in Figure 53a, which shows a schematic cross section of the cell. The cross-bar is very thin and narrow and therefore has a low critical current. When the induced current exceeds this critical value, the cross-bar becomes normal. The flux now changes to the configuration shown in Figure 53b, as the remainder of the film remains superconducting. If finally the drive current is again removed, the superconductivity of the cross-bar is restored, and now the flux threading the hole is trapped, as long as the

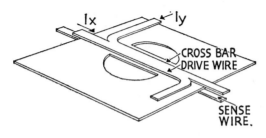

Fig. 52

cross-bar remains superconducting, by a persistent current which is in the opposite direction of the originally induced flow. Even when the drive wire current is now removed, the flux distribution remains that of Figure 53c.

The idealized operation of a Crowe cell (Garwin, 1957) is indicated in Figure 54, which shows on equal time scales, but arbitrary vertical scales, the drive current $I_d$, and the cross-bar current $I_c$. Pulse 1 is too small to induce a critical value of $I_c$. Pulse 2 results in $I_c > I_{crit}$; the cross-bar becomes momentarily normal, and after the drive pulse is removed a persistent current $I_p$ is stored. Pulse 3 is now a 'read-out' pulse which has no effect since it induces a current in a direction opposite to that of the persistent current. With pulse 4, however, the persistent current is reversed, storing the other possibility of the two-state memory, and now read-out pulse 5 succeeds in driving the cross-bar well beyond the critical value. Note that this is a destructive read-out.

The memory is sensed by means of a wire below the cross-bar, also very close to it but electrically insulated. A current pulse will be induced in the sense wire because of its proximity whenever the flux linking the cross-bar changes, that is, whenever the cross-bar becomes normal. Thus we note on Figure 53 that the sense wire response $I_s$ to pulse 3 is nothing, which can be taken as 'Read 0', while its response to 5 is a pulse which can be taken as 'Read 1'.

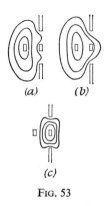

(a)    (b)

(c)

FIG. 53

The operation of the Crowe cell is rendered more complicated than is indicated in the preceding simplified account because the cross-bar heats up through joule heat when it becomes normal, and the thermal recovery time may be appreciable. Crowe (1957), Rhoderick (1959), Von Ballmoos (1961), and several papers in [9] discuss the resulting complications.

Crowe cells can be arranged into a two-dimensional matrix of memory elements with the drive wire forming part both of an $x$- and a $y$-circuit, as indicated in Figure 52. Driving pulses $I_x$, $I_y$ are then so chosen that either alone is not sufficient to activate the device, but that both together do. The reader is again referred to [9] for a number of papers on superconducting memories built up of such matrices. Rose-Innes (1959) has estimated the consumption of liquid helium required to keep a memory like that cold, and finds this to be of the

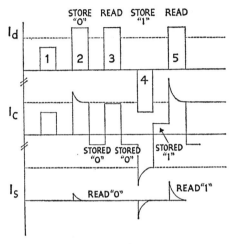

Fig. 54

order of two litres per hour for an array of one million cells. This is well within the capacity of closed cycle helium refrigerators such as the one described by McMahon and Gifford (1960).

# Bibliography

**General References**

[1] SHOENBERG, D., *Superconductivity*, Cambridge University Press, 1952.

[2] LONDON, F., *Superfluids*, Vol. I, New York, Wiley, 1950.

[3] *Progress in Low Temperature Physics*, Vol. I, C. J. Gorter, ed.; New York, Interscience, 1955.

[4] SERIN, B., *Superconductivity: Experimental Part, Handbuch der Physik*, Vol. XV, S. Flügge, ed.; Berlin, Springer, 1956.

[5] BARDEEN, J., *Theory of Superconductivity*, ibid.

[6] *Proc. VII Int. Conf. Low Temp. Phys.*, G. M. Graham and A. C. Hollis-Hallett, eds.; Toronto, University Press, 1960.

[7] BARDEEN, J., and SCHRIEFFER, J. R., *Recent Developments in Superconductivity, Progress in Low Temperature Physics*, Vol. III, C. J. Gorter, ed.; New York, Interscience, 1961.

[8] BARDEEN, J., *Review of the Present Status of the Theory of Superconductivity, IBM Journal* **6**, 3 (1962).

[9] *Proc. Symp. Supercond. Techniques*, Washington, May, 1960; *Solid State Electr.* **1**, 255–408 (1960).

[10] TINKHAM, M., *Superconductivity, Low Temperature Physics*, C. de Witt, B. Dreyfus, and P. G. De Gennes, eds.; London, Gordon and Breach, 1962.

[11] *High Magnetic Fields*, H. Kolm, B. Lax, F. Bitter, and R. Mills, eds.; New York, Wiley, 1962.

[12] *Proc. VIII Int. Conf. Low Temp. Phys.*, R. Davies, ed.; London, Butterworth, 1964.

[13] *Proc. Int. Conf. on the Science of Superconductivity*, Colgate, 1963; published in *Rev. Mod. Physics*, **36**, 1–331 (1964).

[14] DE GENNES, P. G., *Superconductivity of Metals and Alloys*, New York, W. A. Benjamin, Inc., 1966.

[15] *Proc. Tenth Int. Conf. Low Temp. Phys.*, M. P. Malcov, ed., Moscow, 1967.

[16] *Superconductivity*, R. D. Parks, ed., New York, Marcel Dekker, Inc., 1968.

**Individual References**

ABELES, B., and GOLDSTEIN, Y. (1965), *Phys. Rev. Lett.* **14**, 256; see also COHEN, R. W., and ABELES, B., [15], Vol. IIB, p. 297.

ABRAHAMS, E., and TSUNETO, T. (1966), *Phys. Rev.* **152**, 416.

ABRAHAMS, E., and WEISS, P. R. (1959), *Cambridge Conference on Superconductivity* (unpublished).

ABRIKOSOV, A. A. (1957), *J.E.T.P. USSR* **32**, 1442; *Soviet Phys. J.E.T.P.* **5**, 1174; *J. Phys. Chem. Solids* **2**, 199.

ABRIKOSOV, A. A., and GORKOV, L. P. (1960), *J.E.T.P. USSR* **39**, 1781; *Soviet Phys. J.E.T.P.* **12**, 1243 (1961).

ABRIKOSOV, A. A., GORKOV, L. P., and KHALATNIKOV, I. M. (1958), *J.E.T.P. USSR* **35**, 265; *Soviet Phys. J.E.T.P.* **8**, 182 (1959); see also KHALATNIKOV, I. M., and ABRIKOSOV, A. A. (1959), *Advances in Physics*, N. F. Mott, ed., London, Taylor and Francis Ltd.

ADKINS, C. J. (1964), [13], p. 24.

ADKINS, C. J., and KINGTON, B. W. (1966), *Phil. Mag.* **10**, 971.

ADLER, J. G., JACKSON, J. E., and CHANDRASEKHAR, B. S. (1966), *Phys. Rev. Lett.* **16**, 53.

ADLER, J. G., and ROGERS, J. S. (1963), *Phys. Rev. Lett.* **10**, 217.

ADLER, J. G., ROGERS, J. S., and WOODS, S. B. (1965), *Can. J. Phys.* **43**, 557.

ALEKSEEVSKII, N. E. (1953), *J.E.T.P. USSR* **24**, 240.

ALERS, G. A., and WALDORF, D. L. (1961), *Phys. Rev. Lett.* **6**, 677; *IBM Journal* **6**, 89 (1962).

ALERS, P. B. (1957), *Phys. Rev.* **105**, 104.

ALERS, P. B. (1959), *Phys. Rev.* **116**, 1483.

AMBEGAOKAR, V. (1966), *Phys. Rev. Lett.* **16**, 1047.

AMBEGAOKAR, V., and BARATOFF, A. (1963), *Phys. Rev. Lett.* **10**, 486; Erratum: *ibid.*, **11**, 104.

AMBEGAOKAR, V., and TEWORDT, L. (1964), *Phys. Rev.* **134**, A805.

AMBEGAOKAR, V., and WOO, J. W. F. (1965), *Phys. Rev.* **139**, A1818.

AMBLER, E., COLWELL, J. H., HOSLER, W. R., and SCHOOLEY, J. F. (1966), *Phys. Rev.* **148**, 280.

ANDERSON, P. W. (1959), *J. Phys. Chem. Solids* **11**, 26.

ANDERSON, P. W. (1960), [6], p. 298.

ANDERSON, P. W. (1962), *Phys. Rev. Lett.* **9**, 309.

ANDERSON, P..W. (1964), *Lect. on the Many Body Problem*, Vol. II, E. R. Caianiello, ed.; New York: Academic Press, pp. 113–35.

ANDERSON, P. W., and DAYEM, A. H. (1964), *Phys. Rev. Lett.* **13**, 195.

ANDERSON, P. W., and KIM, Y. B. (1964), [13], p. 39.

ANDERSON, P. W., and ROWELL, J. M. (1963), *Phys. Rev. Lett.* **10**, 230.

ANDERSON, P. W., and SUHL, H. (1959), *Phys. Rev.* **116**, 898.

ANDREEV, A. F. (1964), *J. E. T. P. USSR* **46**, 1823; *Soviet Phys. J. E. T. P.* **19**, 1228.

ANDRES, K., OLSEN, J. L., and ROHRER, H. (1962), *IBM Journal* **6**, 84.

ANDREWS, D. H., MILTON, R. M., and DE SORBO, W. (1946), *J. Opt. Soc.* **36**, 518.

ANDROES, G. M., and KNIGHT, W. D. (1961), *Phys. Rev.* **121**, 779.

AOKI, R., and OHTSUKA, T. (1965), *Phys. Rev. Lett.* **19**, 456; [15] (1966).

APPEL, J. (1965), *Phys. Rev.* **139**, A1536.

APPEL, J. (1967), *Phys. Rev.* **156**, 421.

AUTLER, S. H. (1960), *Rev. Sci. Inst.* **31**, 369.

AUTLER, S. H., ROSENBLUM, E. S., and GOOEN, K. H. (1962), *Phys. Rev. Lett.* **9**, 489; see also [13], p. 77.

BALASHOVA, B. M., and SHARVIN, YU. V. (1947), *J.E.T.P. USSR* **17**, 851.

BALLMOOS, F. VON (1961), Thesis, E.T.H., Zürich.

BALTENSPERGER, K. (1959), *Helv. Phys. Acta* **32**, 197.

BARDASIS, A., and SCHRIEFFER, J. R. (1961), *Phys. Rev. Lett.* **7**, 79.

BARDEEN, J. (1950), *Phys. Rev.* **80**, 567; see also *Rev. Mod. Phys.* **23**, 261 (1951).

BARDEEN, J. (1952), *Phys. Rev.* **87**, 192.

BARDEEN, J. (1954), *Phys. Rev.* **94**, 554.

BARDEEN, J. (1957), *Nuovo Cimento* **5**, 1766.

BARDEEN, J. (1958), *Phys. Rev. Lett.* **1**, 399.

BARDEEN, J. (1959), *Cambridge Conference on Superconductivity* (unpublished).

BARDEEN, J. (1961a), *Phys. Rev. Lett.* **6**, 57.

BARDEEN, J. (1961b), *Phys. Rev. Lett.* **7**, 162.

BARDEEN, J. (1962a), *Phys. Rev. Lett.* **9**, 147.

BARDEEN, J. (1962b), *Rev. Mod. Phys.* **34**, 667.

BARDEEN, J., COOPER, L. N., and SCHRIEFFER, J. R. (1957), *Phys. Rev.* **108**, 1175.

BARDEEN, J., RICKAYZEN, G., and TEWORDT, L. (1959), *Phys. Rev.* **113**, 982.

BARDEEN, J., and STEPHEN, M. J. (1965), *Phys. Rev.* **140**, A1197.

BARNES, L. J., and HAKE, R..R. (1967), *Phys. Rev.* **153**, 435.

BEAN, C. P. (1962), *Phys. Rev. Lett.* **8**, 250.

BEAN, C. P. (1964), *Rev. Mod. Phys.* **36**, 31.

BEAN, C. P., and LIVINGSTON, J. D. (1964), *Phys. Rev. Lett.* **12**, 14.

BECKER, R., HELLER, G., and SAUTER, F. (1933), *Z. Phys.* **85**, 772.

BENNEMAN, K. H., and GARLAND, J. W. (1967), *Phys. Rev.* **159**, 369.

BERGMANN, G. (1965), *Z. Phys.* **192**, 379.

BERLINCOURT, T. G. (1963), *Brit. J. Appl. Phys.* **14**, 749.

BERLINCOURT, T. G., and HAKE, R. R. (1962), *Phys. Rev. Lett.* **9**, 293.

BERLINCOURT, T. G., and HAKE, R. R. (1963), *Phys. Rev.* **131**, 140.

BERLINCOURT, T. G., HAKE, R. R., and LESLIE, D. H. (1961), *Phys. Rev. Lett.* **6**, 671.

BERMON, S., and GINSBERG, D. N. (1964), *Phys. Rev.* **135**, A306.

BETBEDER-MATIBET, O., and NOZIÈRES, P. (1961), *C. R. Acad. Sci.* **252**, 3943.

BEZUGLYI, P. O., GALKIN, A. A., and KAROLYUK, A. P. (1959), *J.E.T.P. USSR* **36**, 1951; *Soviet Phys. J.E.T.P.* **9**, 1388.

BEZUGLYI, R. A., GALKIN, A. A., and KAROLYUK, A. P. (1960), *J.E.T.P. USSR* **39**, 7; *Soviet Phys. J.E.T.P.* **12**, 4.

BIONDI, M. A. *et al.* (1965), *Proc. LT* 9, Plenum Press, p. 387.

BIONDI, M. A., FORRESTER, A. T., GARFUNKEL, M. P., and SATTERTHWAITE, C.B. (1958), *Rev. Mod. Phys.* **30**, 1109.

BIONDI, M. A., and GARFUNKEL, M. P. (1959), *Phys. Rev.* **116**, 853, 862.

BIONDI, M. A., GARFUNKEL, M. P., and MCCOUBREY, A. O. (1957), *Phys. Rev.* **108**, 495.

BIONDI, M. A., GARFUNKEL, M. P., and THOMPSON, W. A. (1964), *Phys. Rev.* **136**, A1471.

BLANC, J., GOODMAN, B. B., KUHN, G., LYNTON, E. A., and WEIL, L. (1960), [6], p. 393.

BLUMBERG, R. H. (1962), *J. Appl. Phys.* **33**, 1822.

BOATO, G., GALLINARO, G., and RIZZUTO, C. (1963), *Phys. Lett.* **5**, 20; [13], p. 164 (1964); *Phys. Rev.* **148**, 353.

BOATO, G., GALLINARO, G., and RIZZUTO, C. (1965), *Solid State Comm.* **3**, 173.

BON-MARDION, G., GOODMAN, B. B., and LACAZE, A. (1962), *Phys. Lett.* **2**, 321; *J. Phys. Chem. Solids* **26**, 1143 (1965).

BOORSE, H. A. (1959), *Phys. Rev. Lett.* **2**, 391.

BORCHERDS, P. H., GOUGH, C. E., VINEN, W. F., and WARREN, A. C. (1964), *Phil. Mag.* **10**, 349.

BOZORTH, R. M., DAVIS, D. O., and WILLIAMS, A. J. (1960), *Phys. Rev.* **119**, 1570.

BRENIG, W. (1961), *Phys. Rev. Lett.* **7**, 337.

BRINK, D. M., and ZUCKERMANN, M. J. (1965), *Proc. Phys. Soc.* **85**, 329.

BROCKHOUSE, B. N., ARASE, T., CAGLIOTI, G., RAO, K. R., and WOODS, A. D. B. (1962), *Phys. Rev.* **128**, 1099.

BROWN, A., ZEMANSKY, M. W., and BOORSE, H. A. (1953), *Phys. Rev.* **92**, 52.

BRYANT, C. A., and KEESOM, P. H. (1960), *Phys. Rev. Lett.* **4**, 460; *Phys. Rev.* **123**, 491 (1961).

BUCHANAN, J., CHANG, G. K., and SERIN, B. (1965), *J. Phys. Chem. Solids*, **26**, 1183.

BUCHER, E., GROSS, D., and OLSEN, J. L. (1961), *Helv. Phys. Acta* **34**, 775.

BUCHER, E., and OLSEN, J. L. (1964), [12], p. 139.

BUCHER, E., MULLER, J., OLSEN, J. L., and PALMY, C. (1965), *Phys. Lett.* **15**, 303.

BUCKINGHAM, M. J. (1958), *Proc. V Int. Conf. Low Temp. Phys., Madison*, J. R. Dillinger, ed.; Madison, U. of Wisconsin Press, p. 229.

BUDNICK, J. I. (1960), *Phys. Rev.* **119**, 1578.

BUDNICK, J. I., LYNTON, E. A., and SERIN, B. (1956), *Phys. Rev.* **103**, 286.

BUDZINSKI, W. V., and GARFUNKEL, M. P. (1966), *Phys. Rev. Lett.* **16**, 1100; **17**, 24.

BURGER, J. P., DEUTSCHER, G., GUYON, E., and MARTINET, A. (1965), *Phys. Lett.* **16**, 220.

BURSTEIN, E., LANGENBERG, D. N., and TAYLOR, B. N. (1961), *Phys. Rev. Lett.* **6**, 92.

BYERS, N., and YANG, C. N. (1961), *Phys. Rev. Lett.* **7**, 46.

CALVERLEY, A., MENDELSSOHN, K., and ROWELL, P. M. (1961), *Cryogenics* **2**, 26.

CALVERLEY, A., and ROSE-INNES, A. C. (1960), *Proc. Roy. Soc.* **A255**, 267.

CAPE, J. A. (1963), *Phys. Rev.* **132**, 1486.

CAPE, J. A. (1966), *Phys. Rev.* **148**, 257.

CAROLI, C., and CYROT, M. (1965), *Phys. Kondens. Materie* **4**, 285.

CAROLI, C., CYROT, M. and DE GENNES, P. G. (1966), *Solid State Commun.* **4**, 17.

CAROLI, C., DE GENNES, P. G., and MATRICON, J. (1962), *J. Phys. Rad.* **23**, 707.

CAROLI, C., DE GENNES, P. G., and MATRICON, J. (1964), *Phys. Lett.* **9**, 307; see also CAROLI, C., and MATRICON, J. (1965), *Phys. Kondens. Materie* **3**, 380.

CAROLI, C., and MAKI, K. (1967), *Phys. Rev.* **164**, 591.

CAROLI, C., and MATRICON, J. (1965), *Phys. Kondens. Materie* **3**, 380.

CASIMIR, H. B. G. (1938), *Physica* **5**, 595.

CASIMIR, H. B. G. (1940), *Physica* **7**, 887.

CHAMBERS, R. G. (1952), *Proc. Roy. Soc.* **A215**, 481.

CHAMBERS, R. G. (1956), *Proc. Camb. Phil. Soc.* **52**, 363.

CHANDRASEKHAR, B. S. (1962), *Phys. Lett.* **1**, 7.

CHANDRASEKHAR, B. S., DINEWITZ, I. J., and FARRELL, D. E. (1966), *Phys. Lett.* **20**, 321.

CHANG, G. K., JONES, R. E., and TOXEN, A. M. (1962), *IBM Journal* **6**, 112.

CHANG, G. K., KINSEL, T., and SERIN, B. (1963), *Phys. Lett.* **5**, 11.

CHANG, G. K., and SERIN, B. (1966), *Phys. Rev.* **145**, 274.

CHANIN, G., LYNTON, E. A., and SERIN, B. (1959), *Phys. Rev.* **114**, 719.

CHIOU, C., QUINN, D., and SERAPHIM, D. (1961), *Bull. Am. Phys. Soc.* **6**, 122; SERAPHIM, D., CHIOU, C., and QUINN, D., *Acta Met.* **9**, 861 (1961).

CLAESON, T., and GYGAX, S. (1966), *Sol. State Comm.* **4**, 385.

CLAESON, T., GYGAX, S., and MAKI, K. (1967), *Phys. Kondens. Materie* **6**, 23.

CLAIBORNE, L. T., and EINSPRUCH, N. G. (1966), *Phys. Rev.* **151**, 229.

CLAIBORNE, L. T., and MORSE, R. W. (1964), *Phys. Rev.* **136**, A893.

CLARKE, J. (1966), *Phil. Mag.* **13**, 115.

CLOGSTON, A. M. (1962), *Phys. Rev. Lett.* **9**, 266.

CLOGSTON, A. M., GOSSARD, A. C., JACCARINO, V., and YAFET, Y. (1962), *Phys. Rev. Lett.* **9**, 262; see also [13], p. 170.

COCHRAN, J. F., MAPOTHER, D. E., and MOULD, R. E. (1958), *Phys. Rev.* **103**, 1657.

CODY, G. D. (1958), *Phys. Rev.* **111**, 1078.

COFFEY, H. T., FOX, D. K., HULM, J. K., SPAN, R. E., and REYNOLDS, W. T. (1964), *Bull. Am. Phys. Soc.* **9**, 454.

COHEN, M. H., FALICOV, L. M., and PHILLIPS, J. C. (1962), *Phys. Rev. Lett.* **8**, 316; see also [12], p. 178.

COHEN, M. L. (1964), [13], p. 240; *Phys. Rev.* **134**, A511.

COLLIER, R. S., and KAMPER, R. A. (1966), *Phys. Rev.* **143**, 323.

COLLINS, S. C. (1959) (unpublished). I am grateful to Prof. Collins for a detailed description of this experiment.

COMPTON, V. B., and MATTHIAS, B. T. (1959), *Acta Cryst.* **12**, 651.

COMPTON, W. DALE (1959), private communication to B. Serin.

CONOLLY, A., and MENDELSSOHN, K. (1962), *Proc. Roy. Soc.* **A266**, 429.

COOK, C. F., and EVERETT, G. E. (1967), *Phys. Rev.* **159**, 374.

COOPER, L. N. (1956), *Phys. Rev.* **104**, 1189.

COOPER, L. N. (1959), *Phys. Rev. Lett.* **3**, 17.

COOPER, L. N. (1960), *Am. J. Phys.* **28**, 91.

COOPER, L. N. (1961), *Phys. Rev. Lett.* **6**, 689; *IBM Journal* **6**, 75 (1962).

COOPER, L. N. (1962), *Phys. Rev. Lett.* **8**, 367; see also [12], p. 126.

COOPER, L. N., HOUGHTON, A., and LEE, H. J. (1965), *Phys. Rev. Lett.* **15**, 584.

CORAK, W. S., GOODMAN, B. B., SATTERTHWAITE, C. B., and WEXLER, A. (1954), *Phys. Rev.* **96**, 1442; see also *Phys. Rev.* **102**, 656 (1956).

CORAK, W. S., and SATTERTHWAITE, C. B. (1954), *Phys. Rev.* **99**, 1660.

CORNISH, F. H. J., and OLSEN, J. L. (1953), *Helv. Phys. Acta* **26**, 369.

CRIBIER, D., JACROT, B., MADHOV RAO, L., and FARNOUX, B. (1964), *Phys. Lett.* **9**, 106.

CRITTENDEN, JR., E. C. (1958) *Proc. V Int. Conf. Low Temp. Phys., Madison*, J. R. Dillinger, ed.; Madison, U. of Wisconsin Press, p. 232.

CROW, J. E., and PARKS, R. D. (1966), *Phys. Lett.* **21**, 378.

CROWE, J. W. (1957), *IBM Journal* **1**, 295.

CROWE, J. W. (1958), *Proc. V Int. Conf. Low Temp. Phys., Madison*, J. R. Dillinger, ed.; Madison, U. of Wisconsin Press, p. 238.

CULLEN, J. R., and FERRELL, R. A. (1966), *Phys. Rev.* **146**, 283.

CULLER, G. J., FRIED, B. D., HUFF, R. W., and SCHRIEFFER, J. R. (1962), *Phys. Rev. Lett.* **8**, 399.

CYROT, M. and MAKI, K. (1967), *Phys. Rev.* **156**, 443.

DAUNT, J. G., and MENDELSSOHN, K. (1946), *Proc. Roy. Soc.* **A185**, 225.

DAUNT, J. G., MILLER, A. R., PIPPARD, A. B., and SHOENBERG, D. (1948), *Phys. Rev.* **74**, 842.

DAVIES, E. A. (1960), *Proc. Roy. Soc.* **A255**, 407.

DAYEM, A. H., and MARTIN, R. J. (1962), *Phys. Rev. Lett.* **8**, 246.

DEATON, B. C. (1966), *Phys. Rev. Lett.* **16**, 577.

DEAVER, JR., B. S., and FAIRBANK, W. M. (1961), *Phys. Rev. Lett.* **7**, 43.

DE BLOIS, R. W., and DE SORBO, W. (1964), *Phys. Rev. Lett.* **12**, 499.

DE GENNES, P. G. (1963), *Phys. Lett.* **5**, 22.

DE GENNES, P. G. (1964), [13], p. 225.

DE GENNES, P. G. (1964), *Phys. Kondens. Materie* **3**, 79.

DE GENNES, P. G., and GUYON, E. (1962), *Phys. Lett.* **3**, 168.

DE GENNES, P. G., and MATRICON, J. (1964), [13], p. 45.

DE GENNES, P. G. and MAURO, S. (1965), *Solid State Comm.* **3**, 381.

DE GENNES, P. G., and SARMA, G. (1963), *J. Appl. Phys.* **34**, 1380.

DE GENNES, P. G., and TINKHAM, M. (1964), *Physics* **1**, 107.

DÉSIRANT, M., and SHOENBERG, D. (1948), *Proc. Phys. Soc.* **60**, 413.

DE SORBO, W. (1959), *J. Phys. Chem. Solids* **15**, 7.

DE SORBO, W. (1960), *Phys. Rev. Lett.* **4**, 406; **6**, 369.

DEUTSCHER, G., and DE GENNES, P. G. (1968) [16].

DIETRICH, I. (1964), [12], p. 173.

DOIDGE, P. R. (1956), *Phil. Trans. Roy. Soc.* **A248**, 553.

DOLL, R., and NÄBAUER, M. (1961), *Phys. Rev. Lett.* **7**, 51.

DONALDSON, G. B. (1966), [15], Vol. IIB, p. 291.

DOUGLASS, JR., D. H. (1961a), *Phys. Rev. Lett.* **6**, 346.

DOUGLASS, JR., D. H. (1961b), *Phys. Rev. Lett.* **7**, 14.

DOUGLASS, JR., D. H. (1961c), *Phys. Rev.* **124**, 735.

DOUGLASS, JR., D. H. (1962a), *Bull. Am. Phys. Soc.* **7**, 197.

DOUGLASS, JR., D. H. (1962b), *Phys. Rev. Lett.* **9**, 155.

DOUGLASS, JR., D. H., and BLUMBERG, R. H. (1962). *Phys. Rev.* **127**, 2038.

DOUGLASS, JR., D. H., and FALICOV, L. M. (1964), *Progress in Low Temp. Phys.*, Vol. IV, C. J. Gorter, ed.; New York, Interscience.

DOUGLASS, JR., D. H., and MESERVEY, R. H. (1964a), [12], p. 180.

DOUGLASS, JR., D. H., and MESERVEY, R. H. (1964b), *Phys. Rev.* **135**, A19.

DRANGEID, K. E., and SOMMERHALDER, R. (1962), *Phys. Rev. Lett.* **8**, 467.

DRESSELHAUS, M. S., DOUGLASS, JR., D. H., and KYHL, R. L. (1964), [12], p. 328.

DRUYVESTEYN, W. F., VAN GURP, G. J., and GREEBE, C. A. A. J. (1966), *Phys. Lett.* **22**, 248.

DUBECK, L., LINDENFELD, P., LYNTON, E. A., and ROHRER, H. (1963), *Phys. Rev. Lett.* **10**, 98.

DUBECK, L., LINDENFELD, P., LYNTON, E. A., and ROHRER, H. (1964), [13], p. 110.

EBNETH, G., and TEWORDT, L. (1965), *Z. Phys.* **185**, 421.

ECK, R. E., SCALAPINO, D. J., and TAYLOR, B. N. (1964), *Phys. Rev. Lett.* **13**, 15.

EHRENFEST, P. (1933), *Comm. Leiden Suppl.* **75b**.

EILENBERGER, G. (1965), *Z. Phys.* **182**, 427.

EILENBERGER, G. (1966), *Z. Phys.* **190**, 142.

EILENBERGER, G. (1967), *Phys. Rev.* **153**, 584.

ELIASHBERG, G. M. (1960), *J.E.T.P. USSR* **38**, 996; *Soviet Phys. J.E.T.P.* **11**, 696.

ELIASHBERG, G. M. (1962), *J.E.T.P. USSR* **43**, 1105; *Soviet Phys. J.E.T.P.* **16**, 780 (1963).

ERLBACH, E., GARWIN, R. L., and SARACHIK, M. P. (1960), *I.B.M. Journal* **4**, 107.

ESSMANN, U., and TRÄUBLE, H. (1966), *J. Sci. Instr.* **43**, 344.

ESSMANN, U., and TRÄUBLE, H. (1967), *Phys. Lett.* **24A**, 526.

FABER, T. E. (1952), *Proc. Roy. Soc.* **A214**, 392.

FABER, T. E. (1954), *Proc. Roy. Soc.* **A223**, 174.

FABER, T. E. (1955), *Proc. Roy. Soc.* **A231**, 353.

FABER, T. E. (1957), *Proc. Roy. Soc.* **A241**, 531.

FABER, T. E. (1958), *Proc. Roy. Soc.* **A248**, 460.

FABER, T. E. (1961), private communication.

FABER, T. E., and PIPPARD, A. B. (1955a), *Proc. Roy. Soc.* **A231**, 336.

FABER, T. E., and PIPPARD, A. B. (1955b), [3], Chapter IX.

FAGEN, E. A., and GARFUNKEL, M. P. (1967), *Phys. Rev. Lett.* **18**, 897.

FAIRBANK, W., PIERCE, J. M., and WILSON, P. B. (1964), [12], p. 324.

FANELLI, R., and MEISSNER, H. (1966), *Phys. Rev.* **147**, 227, 235.

FASSNACHT, R. E., and DILLINGER, J. R. (1966), *Phys. Rev. Lett.* **17**, 255.

FERREIRA DA SILVA, J., SCHEFFER, J. VAN DUYKEREN, N. W. J., and DOKOUPIL, Z. (1964), *Phys. Lett.* **12**, 166.

FERREIRA DA SILVA, J., VAN DUYKEREN, N. W. J., and DOKOUPIL, Z. (1966), *Physica* **32**, 1253.

FERRELL, R. A. (1961), *Phys. Rev. Lett.* **6**, 541.

FERRELL, R. A., and GLICK, A. J. (1962), *Bull. Am. Phys. Soc.* **7**, 63.

FERRELL, R. A., and GLOVER III, R. E. (1948), *Phys. Rev.* **109**, 1398.

FERRELL, R. A., and PRANGE, R. E. (1963), *Phys. Rev. Lett.* **10**, 479.

FETTER, A. L. (1966), *Phys. Rev.* **147**, 153.

FETTER, A. L., and HOHENBERG, P. C. (1967), *Phys. Rev.* **159**, 330.

FETTER, A. L., HOHENBERG, P. C., and PINCUS, P. (1966), *Phys. Rev.* **147**, 140.

FEUCHT, D. L., and WOODFORD, JR., J. B. (1961), *J. Appl. Phys.* **32**, 1882.

FIBICH, M. (1964), *Bull. Am. Phys. Soc.* **9**, 454.

FINK, H. J. (1966), *Phys. Rev. Lett.* **17**, 696.

FINNEMORE, D. K., HOPKINS, D. C., and PALMER, P. E. (1965a), *Phys. Rev. Lett.* **15**, 891.

FINNEMORE, D. K., JOHNSON, D. L., OSTENSON, J. E., SPEDDING, F. H., and BEAUDRY, B. J. (1965b), *Phys. Rev.* **137**, A550.

FINNEMORE, D. K., and MAPOTHER, D. E. (1962), *Phys. Rev. Lett.* **9**, 288.

FINNEMORE, D. K., STROUSBERG, T. F., and SWENSON, C. A. (1966), *Phys. Rev.* **149**, 231.

FIORY, A. T., and SERIN, B. (1966), *Phys. Rev. Lett.* **16**, 308.

FIORY, A. T., and SERIN, B. (1967a), *Phys. Rev. Lett.* **19**, 227.

FIORY, A. T., and SERIN, B. (1967b), *Phys. Lett.* **25A**, 557.

FISKE, M. D. (1964), [13], p. 221.

FITE II., W., and REDFIELD, A. G. (1966), *Phys. Rev. Lett.* **17**, 381; *Phys. Rev.* **162**, 358 (1967).

FORRESTER, A. T. (1958), *Phys. Rev.* **110**, 776.

FRANZEN, W. (1963), *J. Opt. Soc. Am.* **53**, 596.

FRIEDEL, J., DE GENNES, P. G., and MATRICON, J. (1963), *Appl. Phys. Lett.* **2**, 119.

FRÖHLICH, H. (1950), *Phys. Rev.* **79**, 845.

FULDE, P. (1965), *Phys. Rev.* **137**, A783.

FULDE, P., and MAKI, K. (1965), *Phys. Rev. Lett.* **15**, 675.

FULDE, P., and MAKI, K. (1966a), *Phys. Rev.* **141**, 275; Erratum: *Phys. Rev.* **147**, 414 (1966).

FULDE, P., and MAKI, K. (1966b), *Phys. Kondens. Materie* **5**, 380.

FULDE, P., and STRÄSSLER, S. (1965) *Phys. Rev.* **140**, A519.

GARFUNKEL, M. P., and SERIN, B. (1952), *Phys. Rev.* **85**, 834.

GARLAND, JR., J. W. (1963a), *Phys. Rev. Lett.* **11**, 111.

GARLAND, JR., J. W. (1963b), *Phys. Rev. Lett.* **11**, 114.

GARLAND, JR., J. W. (1964), [12], p. 143.

GARWIN, R. L. (1957), *I.B.M. Journal* **1**, 304.

GASPAROVIC, R. F., TAYLOR, B. N., and ECK, R. E. (1966), *Solid State Comm.* **4**, 59.

GAYLEY, R. I. (1964), *Phys. Lett.* **13**, 278.

GAYLEY, R. I., LYNTON, E. A., and SERIN, B. (1962), *Phys. Rev.* **126**, 43.

GEBALLE, T. H., and MATTHIAS, B. T. (1962), *IBM Journal* **6**, 256.

GEBALLE, T. H., and MATTHIAS, B. T. (1964), [12], p. 159.

GEBALLE, T. H., MATTHIAS, B. T., HULL, JR., G. W., and CORENZWIT, E. (1961), *Phys. Rev. Lett.* **6**, 275.

GEILIKMAN, B. T. (1958), *J.E.T.P. USSR* **34**, 1042; *Soviet Phys. J.E.T.P.* **7**, 721.

GEILIKMAN, B. T., and KRESIN, V. Z. (1958), *Dokl. Akad. nauk USSR* **123**, 259; *Soviet Phys. Doklady* **3**, 1161.

GEILIKMAN, B. T., and KRESIN, V. Z. (1959), *J.E.T.P. USSR* **36**, 959; *Soviet Phys. J.E.T.P.* **9**, 677.

GEISER, R., and GOODMAN, B. B. (1963), *Phys. Lett.* **5**, 30.

GIAEVER, I. (1960a), *Phys. Rev. Lett.* **5**, 147.

GIAEVER, I. (1960b), *Phys. Rev. Lett.* **5**, 464.

GIAEVER, I. (1964), [12], p. 171.

GIAEVER, I., HART, JR., H. R., and MEGERLE, K. (1962), *Phys. Rev.* **126**, 941.

GIAEVER, I., and MEGERLE, K. (1961), *Phys. Rev.* **122**, 1101.

GIBSON, J. W., and HEIN, R. A. (1966), *Phys. Rev.* **141**, 407.

GINSBERG, D. M. (1962), *Phys. Rev. Lett.* **8**, 204.

GINSBERG, D. M. (1964), *Phys. Rev.* **136**, A1167.

GINSBERG, D. M. (1965), *Phys. Rev.* **138**, A1409.

GINSBERG, D. M., RICHARDS, P. L., and TINKHAM, M. (1959), *Phys. Rev. Lett.* **3**, 337.

GINSBERG, D. M., and TINKHAM, M. (1960), *Phys. Rev.* **118**, 990.

GINZBURG, V. L. (1944), *J.E.T.P. USSR* **14**, 134.

GINZBURG, V. L. (1945), *J. Phys. USSR* **9**, 305.

GINZBURG, V. L. (1955), *J.E.T.P. USSR* **29**, 748; *Soviet Phys. J.E.T.P.* **2**, 589 (1956).

GINZBURG, V. L. (1956a), *J.E.T.P. USSR* **30**, 593; *Soviet Phys. J.E.T.P.* **3**, 621; *Dokl. Akad. nauk USSR* **110**, 368; *Soviet Phys. Doklady* **1**, 541.

GINZBURG, V. L. (1956b), *J.E.T.P. USSR* **31**, 541; *Soviet Phys. J.E.T.P.* **4**, 594.

GINZBURG, V. L. (1958a), *J.E.T.P. USSR* **34**, 113; *Soviet Phys. J.E.T.P.* **7**, 78.

GINZBURG, V. L. (1958b), *Physica* **24**, S42.

GINZBURG, V. L., and LANDAU, L. D. (1950), *J.E.T.P. USSR* **20**, 1064; see also GINZBURG, V. L., *Nuovo Cimento* **2**, 1234 (1955).

GLOVER III, R. E., and TINKHAM, M. (1957), *Phys. Rev.* **108**, 243.

GOEDEMOED, S. H., VAN DER GIESSEN, A., DE KLERK, D., and GORTER, C. J. (1963), *Phys. Lett.* **3**, 250.

GOLDSTEIN, Y., ABELES, B., and COHEN, R. W. (1966), *Phys. Rev.* **151**, 349.

GOODMAN, B. B. (1961), *Phys. Rev. Lett.* **6**, 597; *IBM Journal* **6**, 63.

GOODMAN, B. B. (1962a), *IBM Journal* **6**, 62.

GOODMAN, B. B. (1962b), *Phys. Lett.* **1**, 215.

GOODMAN, B. B. (1962c), *J. Phys. Rad.* **23**, 704.

GOODMAN, B. B. (1964a), [13], p. 12.

GOODMAN, B. B. (1964b), *C. R. Acad. Sci.* **258**, 5175.

GOODMAN, B. B., HILLAIRET, J., VEYSSIÉ, J. J., and WEIL, L. (1960), [6], p. 350.

GOREE, W. S., and EDELSACK, E. A. (1967), *Superconducting Devices, A Literature Survey*, Office Naval Research, Washington, D. C.

GOR'KOV, L. P. (1958), *J.E.T.P. USSR* **34**, 735; *Soviet Phys. J.E.T.P.* **7**, 505.

GOR'KOV, L. P. (1959), *J.E.T.P. USSR* **36**, 1918; **37**, 833, 1407; *Soviet Phys. J.E.T.P.* **9**, 1364; **10**, 593, 998 (1960).

GOR'KOV, L. P. (1960), [6], p. 315.

GOR'KOV, L. P., and RUSINOV, A. I. (1964), *J.E.T.P. USSR* **46**, 1363; *Soviet Phys. J.E.T.P.* **19**, 922.

GORTER, C. J. (1933), *Arch. Mus. Teyler* **7**, 378.

GORTER, C. J. (1935), *Physica* **2**, 449.

GORTER, C. J. (1962a), *Phys. Lett.* **1**, 69.

GORTER, C. J. (1962b), *Phys. Lett.* **2**, 26.

GORTER, C. J. (1964), [13], p. 27.

GORTER, C. J., and CASIMIR, H. B. G. (1934a), *Physica* **1**, 306.

GORTER, C. J., and CASIMIR, H. B. G. (1934b), *Phys. Z.* **35**, 963; *Z. techn. Phys.* **15**, 539.

GORTER, C. J., VAN BEELEN, H., and DE BRUYN OUBOTER, R. (1964), *Phys. Lett.* **8**, 13.

GOTTLIEB, M., JONES, C. K., and GARBUNY, M. (1967), *Phys. Lett.* **25A**, 107.

GRAHAM, G. M. (1958), *Proc. Roy. Soc.* **A248**, 522.

GRASSMANN, P. (1936), *Phys. Z.* **37**, 569.

GRAYSON SMITH, H., MANN, K. C., and WILHELM, J. O. (1936), *Trans. Roy. Soc. Can.* **30**, 13.

GRAYSON SMITH, H., and TARR, F. G. A. (1935), *Trans. Roy. Soc. Can.* **29**, 23.

GRIFFIN, A. (1965), *Phys. Rev. Lett.* **15**, 703.

GROFF, R. P., and PARKS, R. D. (1966), [15], Vol. IIA, p. 216; *Phys. Lett.* **22**, 19.

GUÉNAULT, A. M. (1960), [6], p. 409; *Proc. Roy. Soc.* **A262**, 420 (1961).

GUYON, E. (1966), *Adv. Physics* **15**, 417.

GUYON, E., CAROLI, C., and MARTINET, A. (1964), *J. Physique* **25**, 661.

GUYON, E., MARTINET, A., MATRICON, J., and PINCUS, P. (1965), *Phys. Rev.* **138**, A746.

GUYON, E., MARTINET, A., MAURO, S., and MEUNIER, F. (1966), *Phys. Kondens. Materie* **5**, 123.

GUYON, E., MEUNIER, F., and THOMPSON, R. S. (1967), *Phys. Rev.* **156**, 452.

HAAS, W. J. DE, and VOOGD, J. (1931), *Comm. Leiden* **214c**.

HAENSSLER, F., and RINDERER, L. (1960), [6], p. 375.

HAENSSLER, F., and RINDERER, L. (1965), *Phys. Lett.* **16**, 29; *Helv. Phys. Acta* **38**, 448.

HAENSSLER, F., and RINDERER, L. (1966), [15], Vol. IIB, p. 307.

HAKE, R. R. (1967), *Phys. Rev.* **158**, 356.

HAKE, R. R., and BREMMER, W. G. (1964), *Phys. Rev.* **133**, A179; see also HAKE, R. R., [13], p. 124.

HAKE, R. R., LESLIE, D. H., and BERLINCOURT, T. G. (1962), *Phys. Rev.* **127**, 170.

HAKE, R. R., MAPOTHER, D. E., and DECKER, D. L. (1958), *Phys. Rev.* **118**, 1522.

HAMMOND, R. H., and KELLY, G. M. (1964), [13], p. 185.

HAMMOND, R. H., and KELLY, G. M. (1967), *Phys. Rev. Lett.* **18**, 156.

HARDEN, J. L., and ARP, V. (1963), *Cryogenics* **4**, 105.

HAUSER, J. J. (1962), *Phys. Rev. Lett.* **9**, 423.

HAUSER, J. J. (1966), *Phys. Rev. Lett.* **17**, 921.

HAUSER, J. J. (1966), *Physics*, **2**, 247.

HAUSER, J. J., and BUEHLER, E. (1961), *Phys. Rev.* **125**, 142.

HAUSER, J. J., and HELFAND, E. (1962), *Phys. Rev.* **127**, 386.

HAUSER, J. J., and THEUERER, H. C. (1965), *Phys. Lett.* **14**, 270.

HAUSER, J. J., THEUERER, H. C., and WERTHAMER, N. R. (1964), *Phys. Rev.* **136**, A637.

HAUSER, J. J., THEUERER, H. C., and WERTHAMER, N. R. (1966), *Phys. Rev.* **142**, 118.

HAYNES, M. K. (1960), [9], p. 399.

HEATON, J. W., and ROSE-INNES, A. C. (1963), *Appl. Phys. Lett.* **2**, 196.

HEBEL, L. C. (1959), *Phys. Rev.* **116**, 79.

HEBEL, L. C., and SLICHTER, C. P. (1959), *Phys. Rev.* **113**, 1504.

HEER, C. V., and DAUNT, J. G. (1949), *Phys. Rev.* **76**, 854.

HEIN, R. A., FALGE JR., R. L., MATTHIAS, B. T., and CORENZWIT, E. (1959), *Phys. Rev. Lett.* **2**, 500.

HEIN, R. A., and GIBSON, J. W. (1963), *Phys. Rev.* **131**, 1105; **141**, 407 (1966).

HEIN, R. A., GIBSON, J. W., MATTHIAS, B. T., GEBALLE, T. H., and COREN-ZWIT, E. (1962), *Phys. Rev. Lett.* **8**, 313.

HEIN, R. A., GIBSON, J. W., MAZELSKY, R., MILLER, R. C., and HULM, J. K. (1964), *Phys. Rev. Lett.* **12**, 320.

HEIN, R. A., GIBSON, J. W., PABLO, M. R., and BLAUGHER, R. D. (1963), *Phys. Rev.* **129**, 136.

HELFAND, E., and WERTHAMER, N. R. (1964), *Phys. Rev. Lett.* **13**, 686.

HELFAND, E., and WERTHAMER, N. R. (1966), *Phys. Rev.* **147**, 288.

HEMPSTEAD, C. F., and KIM, Y. B. (1963), *Phys. Rev. Lett.* **12**, 145.

HERRING, C. (1958), *Physica* **24**, S184.

HILSCH, P., and HILSCH, R. (1961), *Naturwiss.* **48**, 549.

HIRSHFELD, A. T., LEUPOLD, H. A., and BOORSE, H. A. (1962), *Phys. Rev.* **127**, 1501.

HOHENBERG, P. C. (1963), *J.E.T.P. USSR* **45**, 1208; *Soviet Phys. J.E.T.P.* **18**, 834 (1964).

HOHENBERG, P. C., and WERTHAMER, N. R. (1967), *Phys. Rev.* **153**, 493.

HULM, J. K. (1950), *Proc. Roy. Soc.* **A204**, 98.

ITTNER III, W. B. (1960a), *Phys. Rev.* **119**, 1591.

ITTNER III, W. B. (1960b), *Solid State Jr.*, July/Aug.

JÄGGI, R., and SOMMERHALDER, R. (1959), *Helv. Phys. Acta* **32**, 313.

JÄGGI, R., and SOMMERHALDER, R. (1960), *Helv. Phys. Acta* **33**, 1.

JAKLEVIC, R. C., LAMBE, J., SILVER, A. H., and MERCEREAU, J. E. (1964), *Phys. Rev. Lett.* **12**, 503, 514.

JOINER, W. C. H., and BLAUGHER, R. D. (1964), *Rev. Mod. Phys.* **36**, 67.

JOINER, W. C. H., and SERIN, B. (1961), private communication; see also SERIN (1960).

JONES, R. E., and TOXEN, A. M. (1960), *Phys. Rev.* **120**, 1167.

JOSEPH, A. S., and TOMASCH, W. J. (1964), *Phys. Rev. Lett.* **12**, 219.

JOSEPHSON, B. D. (1962), *Phys. Lett.* **1**, 251.

JOSEPHSON, B. D. (1964), [13], p. 216.

JURANEK, H. J., NEUMANN, L., and TEWORDT, L. (1966), *Z. Phys.* **173**, 459.

KACHINSKII, V. N. (1965), *Cryogenics*, **5**, 34.

KADANOFF, L. P., and FALKO, I. I. (1964), *Phys. Rev.* **136**, A1170.

KADANOFF, L. P., and MARTIN, P. C. (1961), *Phys. Rev.* **124**, 670.

KAGIWADA, R., LEVY, M., RUDNICK, I., KAGIWADA, H., and MAKI, K. (1967), *Phys. Rev. Lett.* **18**, 74.

KAMERLINGH ONNES, H. (1911), *Leiden Comm.* **122b, 124c.**

KAMERLINGH ONNES, H. (1913), *Leiden Comm. Suppl.* **34.**

KAMERLINGH ONNES, H., and TUYN, W. (1924), *Leiden Comm. Suppl.* **50a**; see also TUYN, W., *Leiden Comm.* **198** (1929).

KAPLAN, R., NETHERCOT, A. H., and BOORSE, H. A. (1959), *Phys. Rev.* **116**, 270.

KEESOM, W. H. (1924), 4ᵉ *Congr. Phys. Solvay*, p. 288.

KEESOM, W. H., and KAMERLINGH ONNES, H. (1924), *Comm. Leiden* **174b.**

KELLER, J. B., and ZUMINO, B. (1961), *Phys. Rev. Lett.* **7**, 164.

KHAIKIN, M. S. (1958), *J.E.T.P. USSR* **34**, 1389; *Soviet Phys. J.E.T.P.* **6**, 735.

KHALATNIKOV, I. M. (1959), *J.E.T.P. USSR* **36**, 1818; *Soviet Phys. J.E.T.P.* **9**, 1296.

KHUKHAREVA, I. S. (1961), *J.E.T.P. (USSR)* **41**, 728; *Soviet Phys. J.E.T.P.* **14**, 526 (1962).

KHUKHAREVA, I. S. (1962), *J.E.T.P. (USSR)* **43**, 1173; *Soviet Phys. J.E.T.P.* **16**, 828.

KIKOIN, I. K., and GOOBAR, S. V. (1940), *J. Phys. U.S.S.R.* **3**, 333; see also BROER, L. J. F., *Physica* **13**, 473 (1947).

KINSEL, T., LYNTON, E. A., and SERIN, B. (1962), *Phys. Lett.* **3**, 30.

KINSEL, T., LYNTON, E. A., and SERIN, B. (1964), [13], p. 105.

KIM, Y. B., HEMPSTEAD, C. F., and STRNAD, A. R. (1963), *Phys. Rev.* **129**, 528; **131**, 2486.

KIM, Y. B., HEMPSTEAD, C. F., and STRNAD, A. R. (1964), [13], p. 43.

KIM, Y. B., HEMPSTEAD, C. F., and STRNAD, A. R. (1965), *Phys. Rev.* **139**, A1163.

KLEINER, W. H., ROTH, L. M., and AUTLER, S. H. (1964), *Phys. Rev.* **133**, A1226.

KLEINMAN, L. (1963), *Phys. Rev.* **132**, 2484.

KLEINMAN, L., TAYLOR, B. N., and BURSTEIN, E. (1964), [13], p. 208.

KLEMENS, P. G. (1956), Chapter IV, Vol. XIV, *Handbuch der Physik*, S. Flügge, ed.; Springer Verlag, Berlin.

KOK, J. A. (1934), *Physica* **1**, 1103.

KOCH, J. E., and KUO, C. C. (1967), *Phys. Rev.* **164**, 618.

KOCH, J. E., and PINCUS, P. A. (1967), *Phys. Rev. Lett.* **19**, 1044.

KONDO, J. (1963), *Prog. Theor. Phys.* **29**, 1.

KRESIN, V. Z. (1959), *J.E.T.P. USSR* **36**, 1947; *Soviet Phys. J.E.T.P.* **9**, 1385.

KROPSHOT, R. H., and ARP, V. D. (1961), *Cryogenics* **2**, 1.

KULIK, I. O. (1966), *J.E.T.P. USSR*, **50**, 1617; *Soviet Phys. J.E.T.P.* **23**, 1077.

KUNZLER, J. E. (1961a), *Rev. Mod. Phys.* **33**, 501.

KUNZLER, J. E. (1961b), *Conf. High Magn. Fields, Cambridge, Mass.* (to be published).

KUNZLER, J. E., BUEHLER, E., HSU, F. S. L., MATTHIAS, B. T., and WAHL, C. (1961a), *J. Appl. Phys.* **32**, 325.

KUNZLER, J. E., BUEHLER, E., HSU, F. S. L., and WERNICK, J. E. (1961b), *Phys. Rev. Lett.* **6**, 89.

KUPER, C. G. (1951), *Phil. Mag.* **42**, 961.

LALEVIC, B. (1966), *J. Appl. Phys.* **31**, 1234.

LAMBE, J., SILVER, A. H., MERCEREAU, J. E., and JAKLEVIC, R. C. (1964), *Phys. Lett.* **11**, 15.

LANDAU, L. D. (1937), *J.E.T.P. USSR* **7**, 371; *Phys. Z. Sowjet.* **11**, 129.

LANDAU, L. D. (1943), *J.E.T.P. USSR* **13**, 377.

LANDAU, L. D., and LIFSHITZ, E. M. (1958), *Statistical Physics*, pp. 434 ff.; London, Pergamon Press.

LANGENBERG, D. N., SCALAPINO, D. J., and TAYLOR, B. N. (1965), *Phys. Rev. Lett.* **15**, 294, 842.

LANGENBERG, D. N., SCALAPINO, D. J., and TAYLOR, B. N. (1966), *Proc. I.E.E.E.* **54**, 560.

LAREDO, S. J., and PIPPARD, A. B. (1955), *Proc. Cambridge Phil. Soc.* **51**, 369.

LARKIN, A. I. (1964), *J.E.T.P. USSR* **46**, 2188; *Soviet Phys. J.E.T.P.* **19**, 1478

LASHER, G. (1967), *Phys. Rev.* **154**, 345.

LAUE, M. VON (1949), *Theorie der Supraleitung*, 2nd ed., Berlin, Springer Verlag; Eng. transl. by L. Meyer and W. Band, New York (1952).

LAURMANN, E., and SHOENBERG, D. (1949), *Proc. Roy. Soc.* **A198**, 560.

LAURMANN, E., and SHOENBERG, D. (1947), *Nature* **160**, 747.

LAVERICK, C. (1967), Superconducting Magnet Technology, *Adv. Electr. and Electron Physics*, L. Marton, ed.; New York, Academic Press.

LAX, E., and VERNON, F. L. (1965), *Phys. Rev. Lett.* **14**, 256; see also [15], Vol. IIB, p. 302.

14

LAZAREV, B. G., and SUDOVSTOV, A. I. (1949), *Dokl. Adak. nauk USSR* **69**, 345.

LEIBOWITZ, J. R. (1964), *Phys. Rev.* **133**, A84.

LESLIE, J. D., CAPELLETTI, R. L., GINSBERG, D. M., FINNEMORE, D. K., SPEDDING, F. H., and BEAUDRY, B. J. (1964), *Phys. Rev.* **134**, A309.

LESLIE, J. D., and GINSBERG, D. M. (1964), *Phys. Rev.* **133**, A362.

LEVINE, J. L. (1967), *Phys. Rev.* **155**, 373.

LIFSHITZ, E. M., and SHARVIN, YU. V. (1951), *Dokl. Akad. nauk USSR* **79**, 783.

LINDENFELD, P. (1961), *Phys. Rev. Lett.* **6**, 613.

LINDENFELD, P., LYNTON, E. A., and SOULEN, R. (1966), [15], Vol. IIA, p. 396.

LINDENFELD, P., and MCCONNELL, R. D. (1966), [15], Vol. IIA, p. 397.

LITTLE, W. A. (1967), *Phys. Rev.* **156**, 396.

LITTLE, W. A., and PARKS, R. D. (1962), *Phys. Rev. Lett.* **9**, 9; see also PARKS, R. D., and LITTLE, W. A., *Phys. Rev.* **133**, A97 (1964).

LIU, S. H. (1965), *Phys. Rev.* **137**, A1209.

LIVINGSTON, J. B. (1963), *Phys. Rev.* **129**, 1943; see also *J. Appl. Phys.* **34**, 3028.

LIVINGSTON, J. B. (1964), [13], p. 54.

LOCK, J. M. (1951), *Proc. Roy. Soc.* **A208**, 391.

LOCK, J. M. (1961a), *Cryogenics* **1**, 243.

LOCK, J. M. (1961b), *Cryogenics* **2**, 65.

LOCK, J. M., PIPPARD, A. B., and SHOENBERG, D. (1951), *Proc. Camb. Phil. Soc.* **47**, 811.

LONDON, F. (1936), *Physica* **3**, 450.

LONDON, F., and LONDON, H. (1935a), *Proc. Roy. Soc.* **A149**, 71.

LONDON, F., and LONDON, H. (1935b), *Physica* **2**, 341.

LONDON, H. (1935), *Proc. Roy. Soc.* **A152**, 650.

LONDON, H. (1940), *Proc. Roy. Soc.* **A176**, 522.

LOW, F. J., and HOFFMAN, A. R. (1963), *Appl. Optics* **2**, 649.

LUCAS, G., and STEPHEN, M. J. (1967), *Phys. Rev.* **154**, 349.

LUTES, O. S. (1957), *Phys. Rev.* **105**, 1451.

LYNTON, E. A., and MCLACHLAN, D. (1962), *Phys. Rev.* **126**, 40.

LYNTON, E. A., and SERIN, B. (1958), *Phys. Rev.* **112**, 70.

LYNTON, E. A., SERIN, B., and ZUCKER, M. (1959), *J. Phys. Chem. Solids* **3**, 165.

MCCONVILLE, T., and SERIN, B. (1965a), *Phys. Rev.* **140**, A868.

MCCONVILLE, T., and SERIN, B. (1965b), *Phys. Rev.* **140**, A1169.

MCMAHON, H. O., and GIFFORD, W. E. (1960), [9], p. 273.

MCMILLAN, W. L., and ROWELL, J. M. (1965), *Phys. Rev. Lett.* **14**, 108; see also [15].

MACMILLAN, W. (1966), see Hauser (1966).

MAKI, K. (1963a), *Prog. Theor. Phys.* **29**, 603.

MAKI, K. (1963b), *Prog. Theor. Phys.* **29**, 333.

MAKI, K. (1964a), *Physics* **1**, 21.

MAKI, K. (1964b), *Physics* **1**, 127.

MAKI, K. (1964c), *Phys. Rev. Lett.* **14**, 98.

MAKI, K. (1964d), *Prog. Theor. Phys.* **31**, 731.

MAKI, K. (1965), *Ann. Phys. (New York)* **34**, 363.

MAKI, K. (1966a), *Phys. Rev.* **148**, 362.

MAKI, K. (1966b), *Phys. Rev.* **148**, 370.

MAKI, K. (1967a), *Phys. Rev.* **153**, 428.

MAKI, K. (1967b), *Phys. Rev.* **156**, 437.

MAKI, K. (1967c), *Phys. Rev.* **158**, 397.

MAKI, K. (1968), [16].

MAKI, K., and TSUNETO, T. (1962), *Prog. Theor. Phys.* **28**, 163.

MAKI, K., and TSUZUKI, T. (1965), *Phys. Rev.* **139**, A868.

MAPOTHER, D. E. (1959), *Conf. on Superconductivity, Cambridge* (unpublished).

MAPOTHER, D. E. (1962), *IBM Journal* **6**, 77; see also *Phys. Rev.* **126**, 2021.

MARKOWITZ, D., and KADANOFF, L. P. (1963), *Phys. Rev.* **131**, 563.

MASUDA, Y. (1962a), *IBM Journal* **6**, 24.

MASUDA, Y. (1962b), *Phys. Rev.* **126**, 127.

MASUDA, Y., and REDFIELD, A. G. (1960a), *Bull. Am. Phys. Soc.* **5**, 176; *Phys. Rev.* **125**, 159 (1962).

MASUDA, Y., and REDFIELD, A. G. (1960b), [6], p. 412; see also MASUDA, Y., *Phys. Rev.* **126**, 1271 (1962).

MATRICON, J. (1964), *Phys. Lett.* **9**, 289.

MATTHIAS, B. T. (1957), Chapter V, *Progress Low Temperature Physics*, Vol. II, C. J. Gorter, ed.; New York, Interscience.

MATTHIAS, B. (1960), *J. Appl. Phys.* **31**, 23S.

MATTHIAS, B. T. (1961), *Rev. Mod. Phys.* **33**, 499.

MATTHIAS, B. T. (1962), *IBM Journal* **6**, 250.

MATTHIAS, B. T., and CORENZWIT, E. (1955), *Phys. Rev.* **100**, 626.

MATTHIAS, B. T., CORENZWIT, E., and ZACHARIASEN, W. H. (1958a), *Phys. Rev.* **112**, 89.

MATTHIAS, B. T., COMPTON, V. B., SUHL, H., and CORENZWIT, E. (1959b), *Phys. Rev.* **115**, 1597.

MATTHIAS, B. T., GEBALLE, T. H., COMPTON, V. B., CORENZWIT, E., and HULL, JR., G. W. (1962), *Phys. Rev.* **128**, 588.

MATTHIAS, B. T., GEBALLE, T. H., CORENZWIT, E., and HULL, JR., G. W. (1963), *Phys. Rev.* **129**, 1025.

MATTHIAS, B. T., GEBALLE, T. H., LONGINOTTI, L. D., CORENZWIT, E., HULL, G. W., WILLENS, R. H., and MAITA, J. P. (1967), *Science* **156**, 645.

MATTHIAS, B. T., PETER, M., WILLIAMS, H. J., CLOGSTON, A. M., CORENZWIT, E., and SHERWOOD, R. C. (1960), *Phys. Rev. Lett.* **5**, 542; CLOGSTON, A. M., *et al.*, *Phys. Rev.* **125**, 541 (1962).

MATTHIAS, B. T., SUHL, H., and CORENZWIT, E. (1958b), *Phys. Rev. Lett.* **1**, 92.

MATTHIAS, B. T., SUHL, H., and CORENZWIT, E. (1958c), *Phys. Rev. Lett.* **1**, 449.

MATTHIAS, B. T., SUHL, H., and CORENZWIT, E. (1959a), *J. Phys. Chem. Solids* **13**, 156.

MATTIS, D. C., and BARDEEN, J. (1958), *Phys. Rev.* **111**, 412.

MAXFIELD, B. W. (1967), *Solid State Comm.* **5**, 585.

MAXFIELD, B. W., and MCLEAN, W. L. (1965), *Phys. Rev.* **139**, A1515.

MAXWELL, E. (1950), *Phys. Rev.* **78**, 477.

MAXWELL, E. (1952a), *Phys. Rev.* **86**, 235.

MAXWELL, E. (1952b), *Phys. Today* **5**, No. 12, p. 14.

MAXWELL, E. (1960), *Adv. Cryo. Eng.* **6**, 154.

MAXWELL, E., and STRONGIN, M. (1964), *Rev. Mod. Phys.* **36**, 144.

MEISSNER, H. (1960), *Phys. Rev.* **117**, 672.

MEISSNER, W., and OCHSENFELD, R. (1933), *Naturwiss.* **21**, 787.

MELIK-BARKHUDEROV, T. K. (1965), *J.E.T.P. USSR* **47**, 311; *Soviet Phys. JETP* **20**, 208.

MENDELSSOHN, K. (1935), *Proc. Roy. Soc.* **A152**, 34.

MENDELSSOHN, K. (1955), [3], Chapter X.

MENDELSSOHN, K. (1962), *IBM Journal* **6**, 27.

MENDELSSOHN, K., and MOORE, J. R. (1934), *Nature* **133**, 413.

MENDELSSOHN, K. and OLSEN, J. L. (1950), *Proc. Phys. Soc.* **A63**, 2.

MENDELSSOHN, K., and PONTIUS, R. B. (1937), *Phil. Mag.* **24**, 777.

MENDELSSOHN, K., and RENTON, C. A. (1955), *Proc. Roy. Soc.* **A230**, 157.

MENDELSSOHN, K., and SHIFFMAN, C. A. (1959), *Proc. Roy. Soc.* **A255**, 199.

MERCEREAU, J. E., and CRANE, L. T. (1963), *Phys. Rev. Lett.* **11**, 107.

MERRIAM, M. F. (1966), *Phys. Rev.* **144**, 300.

MERRIAM, M. F., HAGEN, J., and LUO, H. L. (1967), *Phys. Rev.* **154**, 424.

MESERVEY, R., and DOUGLASS, JR., D. H. (1964), *Phys. Rev.* **135**, A24.

MESHKOVSKY, A., and SHALNIKOV, A. (1947), *J.E.T.P. USSR* **17**, 851; *J. Phys. USSR* **11**, 1.

MILLER, P. B. (1960), *Phys. Rev.* **118**, 928.

MILLSTEIN, J., and TINKHAM, M. (1967), *Phys. Rev.* **158**, 325.

MISENER, A. D., and WILHELM, J. O. (1935), *Trans. Roy. Soc. Can.* **29**, 5.

MOCHEL, J. M., and PARKS, R. D. (1966), *Phys. Rev. Lett.* **16**, 1156.

MOORMANN, W. (1967), *Z. Phys.*

MOREL, P. (1959), *J. Phys. Chem. Solids* **10**, 277.

MOREL, P., and ANDERSON, P. W. (1962), *Phys. Rev.* **125**, 1263.

MORIN, J., MAITA, J. P., WILLIAMS, H. J., SHERWOOD, R., WERNICK, J. H., and KUNZLER, J. E. (1962), *Phys. Rev. Lett.* **8**, 275.

MORRIS, D. E., and TINKHAM, M. (1961), *Phys. Rev. Lett.* **6**, 600; see also *Phys. Rev.* **134**, A1154 (1964).

MORSE, R. W., and BOHM, H. V. (1957), *Phys. Rev.* **108**, 1094.

MORSE, R. W., OLSEN, T., and GAVENDA, J. D. (1959), *Phys. Rev. Lett.* **3**, 15; **4**, 193; see also MORSE, R. W., *IBM Journal* **6**, 58 (1962).

MÜHLSCHLEGEL, B. (1959), *Z. Phys.* **155**, 313.

MÜLLER, J. (1959), *Helv. Phys. Acta* **32**, 141.

NAMBU, Y., and TUAN, S. F. (1963), *Phys. Rev. Lett.* **11**, 119; see also *Phys. Rev.* **133**, A1 (1964).

NETHERCOT, A. H. (1961), *Phys. Rev. Lett.* **7**, 226.

NEUMANN, L., and TEWORDT, L. (1966a), *Z. Phys.* **187**, 55.

NEUMANN, L., and TEWORDT, L. (1966b), *Z. Phys.* **191**, 73.

NEURINGER, L. J., and SHAPIRA, Y. (1965), *Phys. Rev.* **140**, A1638.

NEURINGER, L. J., and SHAPIRA, Y. (1966a), *Phys. Rev.* **148**, 231.

NEURINGER, L. J., and SHAPIRA, Y. (1966b), *Phys. Rev. Lett.* **17**, 81.

NICOL, J., SHAPIRO, S., and SMITH, P. H. (1960), *Phys. Rev. Lett.* **5**, 461.

NIESSEN, A. K., and STAAS, F. W. (1965), *Phys. Lett.* **15**, 26.

NOER, R. J., and KNIGHT, W. D. (1964), [13], p. 177.

NOZIÈRES, P., and VINEN, W. F. (1966), *Phil. Mag.* **14**, 667.

OLSEN, J. L. (1958), *Rev. Sci. Inst.* **29**, 537; see also PURCELL, J. R., and PAYNE, E. G. (1960), *Adv. Cryo. Eng.* **6**, 149.

OLSEN, J. L. (1963), *Cryogenics* **2**, 356.

OLSEN, J. L., BUCHER, E., LEVY, M., MULLER, J., CORENZWIT, E., and GEBALLE, T. (1964), [13], p. 168.

OLSEN, J. L., and ROHRER, H. (1957), *Helv. Phys. Acta* **30**, 49.

OLSEN, J. L., and ROHRER, H. (1960), *Helv. Phys. Acta* **33**, 872; see also ANDRES, K., OLSEN, J. L., and ROHRER, H., *IBM Journal* **6**, 84 (1962).

O'NEAL, H. R., and PHILLIPS, N. E. (1965), *Phys. Rev.* **137**, A748.

ONSAGER, L. (1961), *Phys. Rev. Lett.* **7**, 50.

ORSAY GROUP OF SUPERCONDUCTIVITY (1966a), *Physik Kondens. Materie* **5**, 141.

ORSAY GROUP OF SUPERCONDUCTIVITY (1966b), *Quantum Fluids*, D. Brewer, ed., North Holland, Amsterdam.

OTTER, JR., F. A., and SOLOMON, P. R. (1966), *Phys. Rev. Lett.* **16**, 681.

PALMER, L. H., and TINKHAM, M. (1966), [15], Vol. IIB, p. 238.

PARKINSON, D. H. (1962), *Brit. J. Appl. Phys.* **13**, 49.

PARKINSON, D. H. (1964), *Brit. J. Appl. Phys.* **41**, 68.

PARKS, R. D., and GOFF, R. P. (1967), *Phys. Rev. Lett.* **18**, 342.

PARKS, R. D., and MOCHEL, J. M. (1963), *Phys. Rev. Lett.* **11**, 354; see also [13], p. 284.

PARKS, R. D., MOCHEL, J. M., and SURGENT, L. V. (1964), *Phys. Rev. Lett.* **13**, 331a.

PASKIN, A., STRONGIN, M., CRAIG, P. P., and SCHWEITZER, D. G. (1965), *Phys. Rev.* **137**, A1816.

PEARL, J. (1964), *Appl. Phys. Lett.* **5**, 65.

PEIERLS, R. (1936), *Proc. Roy. Soc.* **A155**, 613.

PETER, M. (1958), *Phys. Rev.* **109**, 1857.

PHILLIPS, N. E. (1958), *Phys. Rev. Lett.* **1**, 363.

PHILLIPS, N. E. (1959), *Phys. Rev.* **114**, 676.

PHILLIPS, N. E., and MATTHIAS, B. T. (1960), *Phys. Rev.* **121**, 105.

PICKLESIMER, M. L., and JEKULA, S. T. (1962), *Phys. Rev. Lett.* **9**, 254.

PINCUS, P. (1967), *Phys. Rev.* **158**, 346.

PINES, D. (1958), *Phys. Rev.* **109**, 280.

PIPPARD, A. B. (1947a), *Proc. Roy. Soc.* **A191**, 385.

PIPPARD, A. B. (1947b), *Proc. Roy. Soc.* **A191**, 399.

PIPPARD, A. B. (1948), *Nature* **162**, 68.

PIPPARD, A. B. (1950), *Proc. Roy. Soc.* **A203**, 210.

PIPPARD, A. B. (1951), *Proc. Camb. Phil. Soc.* **47**, 617.

PIPPARD, A. B. (1953), *Proc. Roy. Soc.* **A216**, 547.

PIPPARD, A. B. (1954), Chapter I, *Advances in Electronics and Electron Physics*, L. Marton, ed.; New York, Academic Press.

PIPPARD, A. B. (1955), *Phil. Trans. Roy. Soc.* **A248**, 97.

PIPPARD, A. B. (1960), [6], p. 320.

PIPPARD, A. B., and PULLAN, G. T. (1952), *Proc. Camb. Phil. Soc.* **48**, 188.

POKROVSKII, V. L. (1961), *J.E.T.P. USSR* **40**, 641; *Soviet Phys. J.E.T.P.* **13**, 447.

POKROVSKII, V. L., and RYVKIN, M. S. (1962), *J.E.T.P. USSR* **43**, 92; *Soviet Phys. J.E.T.P.* **16**, 67 (1963).

PRIVOROTSKII, I. A. (1962), *J.E.T.P. USSR* **43**, 133; *Soviet Phys. J.E.T.P.* **16**, 1945 (1963).

QUINN III, D. J., and ITTNER III, W. B. (1962), *J. Appl. Phys.* **33**, 748.

RADEBAUGH, R., and KEESOM, P. H. (1966a), *Phys. Rev.* **149**, 209.

RADEBAUGH, R., and KEESOM, P. H. (1966b), *Phys. Rev.* **149**, 217.

RATTO, C. F., and BLANDIN, A. (1967), *Phys. Rev.* **156**, 513.

RAYFIELD, G. S., and REIF, F. (1963), *Phys. Rev. Lett.* **11**, 305.

REDFIELD, A. G. (1959), *Phys. Rev. Lett.* **3**, 85; see also REDFIELD, A. G., and ANDERSON, A. G. (1959), *Phys. Rev.* **116**, 583.

REED, W A., FAWCETT, E., and KIM, Y. B. (1965), *Phys. Rev. Lett.* **14**, 790.

REESE, W., and STEYERT, JR., W. A. (1962), *Rev. Sci. Inst.* **33**, 43.

REIF, F. (1957), *Phys. Rev.* **106**, 208.

REIF, F., and WOOLF, M. A. (1962), *Phys. Rev. Lett.* **9**, 315; *Phys. Rev.* **137**, A557.

REUTER, G. E. H., and SONDHEIMER, E. H. (1948), *Proc. Roy. Soc.* **A195**, 336.

REYNOLDS, C. A., SERIN, B., and NESBITT, L. B., (1951), *Phys. Rev.* **84**, 691.

REYNOLDS, C. A., SERIN, B., WRIGHT, W. H., and NESBITT, L. B. (1950), *Phys. Rev.* **78**, 487.

RHODERICK, E. H. (1959), *Brit. J. Appl. Phys.* **10**, 193.

RICHARDS, P. L. (1960), [6], p. 333; see also *Phys. Rev.* **126**, 912 (1962).

RICHARDS, P. L. (1961), *Phys. Rev. Lett.* **7**, 412.

RICHARDS, P. L., and TINKHAM, M. (1960), *Phys. Rev.* **119**, 575.

RICKAYSEN, G. (1966), *Proc. Phys. Soc.* **89**, 129.

ROBERTS, B. W. (1966), *Superconductive Materials and Some of Their Properties*, NBS Technical Note 408.

ROSE-INNES, A. C. (1959), *Brit. J. Appl. Phys.* **10**, 452.

ROSE-INNES, A. C., and SERIN, B. (1961), *Phys. Rev. Lett.* **7**, 278.

ROSENBLUM, B., and CARDONA, M. (1964a), *Phys. Lett.* **9**, 220.

ROSENBLUM, B., and CARDONA, M. (1964b), *Phys. Rev. Lett.* **12**, 657.

ROSENBLUM, B., and CARDONA, M. (1964c), *Phys. Lett.* **13**, 33.

ROTHWARF, A., and COHEN, M. (1963), *Phys. Rev.* **130**, 1401.

ROWELL, J. M. (1963), *Phys. Rev. Lett.* **11**, 200.

ROWELL, J. M., ANDERSON, P. W., and THOMAS, D. E. (1963), *Phys. Rev. Lett.* **10**, 334.

ROWELL, J. M., CHYNOWETH, A. G., and PHILLIPS, T. C. (1962), *Phys. Rev. Lett.* **9**, 59.

ROWELL, J. M., and MCMILLAN, W. L. (1966), [15], Vol. IIB, p. 296.

ROWELL, J. M., MCMILLAN, W. L., and ANDERSON, P. W. (1965), *Phys. Rev. Lett.* **14**, 633.

RUEFENACHT, J., and RINDERER, L. (1964), [12], p. 326.

SAINT JAMES, D. (1965), *Phys. Lett.* **16**, 218.

SAINT JAMES, D., and DE GENNES, P. G. (1963), *Phys. Lett.* **7**, 306.

SATTERTHWAITE, C. B. (1960), [6], p. 405; *Phys. Rev.* **125**, 873 (1962).

SCALAPINO, D. J., and ANDERSON, P. W. (1964), *Phys. Rev.* **133**, A921.

SCALAPINO, D. J., SCHRIEFFER, J. R., and WILKINS, J. W. (1966), *Phys. Rev.* **148**, 263,

SCALAPINO, D. J., WADA, Y., and SWIHART, J. C. (1965), *Phys. Rev. Lett.* **14**, 102.

SCHATTKE, W. (1966), *Phys. Lett.* **20**, 245.

SCHAWLOW, A. L. (1956), *Phys. Rev.* **101**, 573.

SCHAWLOW, A. L. (1958), *Phys. Rev.* **109**, 1856.

SCHAWLOW, A. L., and DEVLIN, G. E. (1959), *Phys. Rev.* **113**, 120.

SCHAWLOW, A. L., MATTHIAS, B. T., LEWIS, H. W., and DEVLIN, G. E. (1954), *Phys. Rev.* **95**, 1344.

SCHMID, A. (1966), *Phys. Kondens. Materie* **5**, 302.

SCHOOLEY, J. F., HOSLER, W. R., and COHEN, M. L. (1964), *Phys. Rev. Lett.* **12**, 474.

SCHRIEFFER, J. R. (1957), *Phys. Rev.* **106**, 47.

SCHRIEFFER, J. R. (1961), *IBM Conf. on Superconductivity* (unpublished).

SCHRIEFFER, J. R., and GINSBERG, D. M. (1962), *Phys. Rev. Lett.* **8**, 207.

SCHRIEFFER, J..R., SCALAPINO, D. J., and WILKINS, J. W. (1963), *Phys. Rev. Lett.* **10**, 336.

SCHRIEFFER, J. R., and WILKINS, J. W. (1963), *Phys. Rev. Lett.* **10**, 17.

SCHWETTMAN, H. A., WILSON, P. B., PIERCE, T. M., and FAIRBANK, W. M. (1965), *Int. Adv. Cryo. Eng.*, Vol. 10, Plenum, New York.

SEIDEL, T., and MEISSNER, H. (1966), *Phys. Rev.* **147**, 272.

SERIN, B. (1955), [3], Chapter VII.

SERIN, B. (1960), [6], p. 391.

SERIN, B. (1965), *Phys. Lett.* **16**, 112.

SERIN, B., REYNOLDS, C. A., and LOHMAN, C. (1952), *Phys. Rev.* **86**, 162.

SEVASTYONOV, B. K. (1961), *J.E.T.P. USSR.* **40**, 52; *Soviet Phys. J.E.T.P.* **13**, 35.

SEVASTYONOV, B. K., and SOKOLINA, V. A. (1962), *J.E.T.P. USSR* **42**, 1212; *Soviet Phys. J.E.T.P.* **15**, 840.

SHALNIKOV, A. I., and SHARVIN, YU. V. (1948), *Izv. Akad. nauk USSR* **12**, 195.

SHAPIRO, S. (1963), *Phys. Rev. Lett.* **11**, 80; see also [13], p. 223.

SHAPIRO, S., and JANUS, A. R. (1964), [12], p. 321.

SHAPOVAL, E. A. (1961), *J.E.T.P. USSR* **41**, 877; *Soviet Phys. J.E.T.P.* **14**, 628 (1962).

SHAPOVAL, E. A. (1965), *J.E.T.P. USSR* **49**, 930; *Soviet Phys. J.E.T.P.* **22**, 647, (1966).

SHARVIN, YU. V. (1960), *J.E.T.P. USSR* **38**, 298; *Soviet Phys. J.E.T.P.* **11**, 216.

SHARVIN, YU. V. (1965), *J.E.T.P. Pis'ma* **2**, 287; *J.E.T.P. Lett.* **2**, 183.

SHARVIN, YU. V. (1966), [15].

SHAW, R. W., and MAPOTHER, D. E. (1960), *Phys. Rev.* **118**, 1474.

SHAW, R. W., MAPOTHER, D. E., and HOPKINS, D. C. (1961), *Phys. Rev.* **121**, 86.

SHEN, L. Y. L., SENOZAN, N. M., and PHILLIPS, N. E. (1965), *Phys. Rev. Lett.* **14**, 1025.

SHERRILL, M. D., and EDWARDS, H. H. (1961), *Phys. Rev. Lett.* **6**, 460.

SHIFFMAN, C. A. (1960), [6], p. 373.

SHIFFMAN, C. A. (1961), *IBM Conf. on Superconductivity* (unpublished).

SHOENBERG, D. (1940), *Proc. Roy. Soc.* **A175**, 49.

SHUBNIKOV, L. W., KOTKEVICH, W. I., SHEPELEV, J. D., and RIABININ, J. N. (1937), *J.E.T.P. USSR* **7**, 221.

SILBERNAGEL, B. G., WEGER, M., and WERMICK, J. H. (1966), *Phys. Rev. Lett.* **17**, 384.

SILIN, V. P. (1951), *J.E.T.P. USSR* **21**, 1330.

SILSBEE, F. B. (1916), *J. Wash. Acad. Sci.* **6**, 597.

SILVER, A. H., JAKLEVIC, R. C., and LAMBE, J. (1966), *Phys. Rev.* **141**, 362.

SIMMONS, W. A., and DOUGLASS, JR., D. H. (1962), *Phys. Rev. Lett.* **9**, 153.

SKALSKI, S., BETBEDER-MATIBET, O., and WEISS, P. R. (1964), *Bull. Am. Phys. Soc.* **9**, 30.

SMITH, J. E., and GINSBERG, D. M. (1968), *Phys. Rev.* **167**, 345.

SMITH, P. H., SHAPIRO, S., MILES, J. L., and NICOL, J. (1961), *Phys. Rev. Lett.* **6**, 686.

SPIEWAK, M. (1959), *Phys. Rev.* **113**, 1479.

STAAS, F. W., NIESSEN, A. K., and DRUYVESTEYN, W. F. (1965), *Phys. Lett.* **17**, 231.

STOUT, J. W., and GUTTMAN, L. (1952), *Phys. Rev.* **88**, 703.

STRÄSSLER, S., and WYDER, P. (1963), *Phys. Rev. Lett.* **10**, 225.

STRNAD, A. R., HEMPSTEAD, C. F., and KIM, Y. B. (1964), *Phys. Rev. Lett.* **13**, 794.

STROMBERG, T. F., and SWENSON, C. A. (1962), *Phys. Rev. Lett.* **9**, 370.

STRONGIN, M., PASKIN, A., SCHWEITZER, D. G., KAMMERER, O. F., and CRAIG, P. P. (1964), *Phys. Rev. Lett.* **12**, 442.

15

SUHL, H. (1962), *Low Temperature Physics*, C. De Witt, B. Dreyfus, and P. G. De Gennes, eds.; London, Gordon and Breach.

SUHL, H., and MATTHIAS, B. T. (1959), *Phys. Rev.* **114**, 977.

SUHL, H., MATTHIAS, B. T., and CORENZWIT, E. (1959a), *J. Phys. Chem. Solids* **11**, 347.

SUHL, H., MATTHIAS, B. T., and WALKER, L. R. (1959b), *Phys. Rev. Lett.* **3**, 552.

SUNG, C. C., and SHEN, L. Y. L. (1965), *Phys. Lett.* **19**, 101.

SWARTZ, P. S. (1962), *Phys. Rev. Lett.* **9**, 448.

SWARTZ, P. S., and HART, H. K. (1965), *Phys. Rev.* **137**, A818.

SWENSON, C. A. (1960), *Solid State Physics*, Vol. 11, p. 41, F. Seitz and D. Turnbull, eds.; New York, Academic Press.

SWENSON, C. A. (1962), *IBM Journal* **6**, 82; see also HINRICHS, C. H., and SWENSON, C. A. (1961), *Phys. Rev.* **123**, 1106; and SCHIRBER, J. E., and SWENSON, C. A. (1961), *Phys. Rev.* **123**, 1115.

SWIHART, J. C. (1959), *Phys. Rev.* **116**, 45.

SWIHART, J. C. (1962), *IBM Journal* **6**, 14.

SWIHART, J. C. (1963), *Phys. Rev.* **131**, 73.

SWIHART, J. C. (1966), [15], Vol. IIB, p. 275.

SWIHART, J. C., SCALAPINO, D. J., and WADA, J. (1965), *Phys. Rev. Lett.* **14**, 106.

TAYLOR, B. N., and BURSTEIN, E. (1963), *Phys. Rev. Lett.* **10**, 14.

TAYLOR, B. N., PARKER, W. H., and LANGENBERG, D. N. (1966), [15], Vol. IIA, p. 59.

TEMPLETON, I. M. (1955a), *J. Sci. Inst.* **32**, 172.

TEMPLETON, I. M. (1955b), *J. Sci. Inst.* **32**, 314.

TEWORDT, L. (1962), *Phys. Rev.* **128**, 12.

TEWORDT, L. (1963), *Phys. Rev.* **132**, 595.

TEWORDT, L. (1965a), *Z. Phys.* **18**, 385.

TEWORDT, L. (1965b), *Phys. Rev.* **137**, A1745.

THOMAS, R. L., WU, H. C., and TEPLEY, N. (1966), *Phys. Rev. Lett.* **17**, 22.

THOMPSON, R. S., and BARATOFF, A. (1965), *Phys. Rev. Lett.* **15**, 971.

THOMPSON, R. S., and BARATOFF, A. (1968), *Phys. Rev.* **167**, 361.

THOULESS, D. J. (1960), *Phys. Rev.* **117**, 1256.

TIEN, P. K., and GORDON, J. P. (1963), *Phys. Rev.* **129**, 647.

TINKHAM, M. H. (1958), *Phys. Rev.* **110**, 26.

TINKHAM, M. H. (1962), *IBM Journal* **6**, 49.

TINKHAM, M. H. (1963), *Phys. Rev.* **129**, 2413.

TINKHAM, M. H. (1964a), [13], p. 268.

TINKHAM, M. H. (1964b), *Phys. Lett.* **9**, 217..

TINKHAM, M. H. (1964c), *Phys. Rev. Lett.* **13**, 804.

TINKHAM, M. H., and FERRELL, R. A. (1959), *Phys. Rev. Lett.* **2**, 331.

TOLMACHEV, V. V. (1958): see BOGOLIUBOV, N. N., TOLMACHEV, V. V., and SHIRKOV, D. V., *A new Method in the Theory of Superconductivity*, Section 6.3 (Acad. Sci. USSR Press, Moscow; Translation: Consultants Bureau, Inc., New York, 1959); see also TOLMACHEV, V. V., *Dokl. Akad. nauk USSR* **140**, 563 (1961); *Soviet Phys. Doklady* **6**, 800 (1962).

TOMASH, W. J., and JOSEPH, A. S. (1963), *Phys. Rev. Lett.* **12**, 148.

TOWNSEND, P., and SUTTON, J. (1962), *Phys. Rev.* **128**, 591.

TOXEN, A. M. (1962), *Phys. Rev.* **127**, 382.

TOXEN, A. M., CHANG, G. K., and JONES, R. E. (1962), *Phys. Rev.* **126**, 919.

TSUNETO, T. (1960), *Phys. Rev.* **118**, 1029.

TSUNETO, T. (1962), *Progr. Theor. Phys.* **28**, 857.

TSUZUKI, T. (1964), *Progr. Theor. Phys.* **31**, 388.

VAN BEELEN, H., ARNOLD, A. J. P. T., SYPKENS, H. A., VAN BRAAM HOUCKGEEST, J. P., DE BRUYN OUBOTER, R., BEENAKKER, J. J. M., and TACONIS, K. W. (1965), *Physica* **31**, 413.

VAN DER HOEVEN, JR., B. J. C., and KEESOM, P. H. (1965), *Phys. Rev.* **137**, A103.

VAN OOIJEN, D. J., and VAN GURP, G. J. (1965), *Phys. Lett.* **17**, 230.

VAN VIJFEIJKEN, A. G., and NIESSEN, A. K. (1965), *Phys. Lett.* **16**, 23.

VINEN, W. F., and WARREN, A. C. (1967a), *Proc. Phys. Soc.* **91**, 399.

VINEN, W. F., and WARREN, A. C. (1967b), *Proc. Phys. Soc.* **91**, 409.

VOLGER, J., STAAS, F. A., and VAN VIJFEIJKEN, A. G. (1964), *Phys. Lett.* **9**, 303.

VON MINNIGERODE, G. (1966), *Z. Phys.* **192**, 379.

VROOMEN, A. R. DE (1955), *Conf. Phys. Basses Temp.*, Paris, p.580.

VROOMEN, A. R. DE, and BAARLE, C. VAN (1957), *Physica* **23**, 785.

WALDRAM, J. R. (1964), *Adv. Phys.* **13**, 1.

WALDRAM, J. R. (1966), [15], Vol. IIA, p. 207.

WATSON, J. H. P., and GRAHAM, G. M. (1963), *Can. J. Phys.* **41**, 1738.

WERTHAMER, N. R. (1963a), *Phys. Rev.* **132**, 663.

WERTHAMER, N. R. (1963b), *Phys. Rev.* **132**, 2440.

WERTHAMER, N. R., HELFAND, E., and HOHENBERG, P. C. (1966), *Phys. Rev.* **147**, 295.

WHITEHEAD, C. S. (1956), *Proc. Roy. Soc.* **A238**, 175.

WIPF, S. (1961), thesis, University of London; see also COLES, B. R., *IBM Journal* **6**, 68 (1962).

WIPF, S., and COLES, B. R. (1959), *Cambridge Superconductivity Conference* (unpublished); see also COLES, B. R., *IBM Journal* **6**, 68 (1962).

WOO, J. W. F. (1967), *Phys. Rev.* **155**, 429.

WOOLF, M. A., and REIF, F. (1965), *Phys. Rev.* **137**, A557.

YANSON, I. K., SVISTUNOV, V. M., and DMITRENKO, I. M. (1965), *J.E.T.P. USSR* **48**, 976; **49**, 1741; *Soviet Phys. J.E.T.P.* **21**, 650; **22**, 1190 (1966).

YAQUB, M. (1960), *Cryogenics* **1**, 101, 166.

YNTEMA, G. B. (1955), *Phys. Rev.* **98**, 1197.

YOUNG, D. R. (1959), *Progr. Cryogenics*, Vol. I, p. 1, K. Mendelssohn, ed.; London, Heywood & Co.

ZAVARITSKII, N. V. (1951), *Dokl. Akad. nauk USSR* **78**, 665.

ZAVARITSKII, N. V. (1952), *Dokl. Akad. nauk USSR* **85**, 749.

ZAVARITSKII, N. V. (1958a), *J.E.T.P. USSR* **33**, 1805; *Soviet Phys. J.E.T.P.* **6**, 837.

ZAVARITSKII, N. V. (1958b), *J.E.T.P. USSR* **34**, 1116; *Soviet Phys. J.E.T.P.* **7**, 773.

ZAVARITSKII, N. V. (1959), *J.E.T.P. USSR* **37**, 1506; *Soviet Phys. J.E.T.P.* **10**, 1069.

ZAVARITSKII, N. V. (1960a), *J.E.T.P. USSR* **38**, 1673; *Soviet Phys. J.E.T.P.* **11**, 1207.

ZAVARITSKII, N. V. (1960b), *J.E.T.P. USSR* **39**, 1193; *Soviet Phys. J.E.T.P.* **12**, 831 (1961).

ZAVARITSKII, N. V. (1960c), *J.E.T.P. USSR* **39**, 1571; *Soviet Phys. J.E.T.P.* **12**, 1093 (1961).

ZAVARITSKII, N. V. (1961), *J.E.T.P. USSR* **41**, 657; *Soviet Phys. J.E.T.P.* **14**, 470 (1962).

ZIMMERMAN, J. E., and MERCEREAU, J. E. (1965), *Phys. Rev. Lett.* **14**, 887.

ZIMMERMAN, J. E., and SILVER, A. H. (1966), *Phys. Rev.* **141**, 367.

ZUCKERMANN, M. J. (1965), *Phys. Rev.* **140**, A899.

ZUMINO, B., and UHLENBROCK, D. A. (1964), *Nuovo Cimento* **33**, 1446.

# Index